THE NEW YORK STATE CANAL SYSTEM

THE NEW YORK STATE CANAL SYSTEM

A History Beyond the Erie

SUSAN P. GATELEY

Published by The History Press
Charleston, SC
www.historypress.com

First published 2023

Manufactured in the United States

ISBN 9781467154178

Library of Congress Control Number: 2022951590

Notice: The information in this book is true and complete to the best of our knowledge. It is offered without guarantee on the part of the author or The History Press. The author and The History Press disclaim all liability in connection with the use of this book.

CONTENTS

PREFACE

No state has a greater variety of navigable waterways than New York. It has shoreline on two Great Lakes as well as on a pretty good lake, that of Lake Champlain, and it's the only state with frontage on the ocean and the Great Lakes thanks to its saltwater coast of Long Island Sound and the Atlantic. Within New York's boundaries lie hundreds of small mountain lakes along with white water streams, large rivers and the much-celebrated Finger Lakes of Central New York. And it has the 524-mile-long contemporary state canal system that stretches from Albany to Buffalo and from the southern tier counties of Tompkins and Schuyler to the Quebec border. The navigable canal system follows much of the route of the original Erie and several other nineteenth-century waterways and spans much of the state. It is unique in all America.

Nature dictated the success of the state's original canal. Thanks to the ice age glaciers that covered New York, there are no major highlands between Albany and Lake Erie. With one exception, the major rivers and streams of that corridor run east and west. About fourteen thousand years ago, as the last glacier began to retreat north, it opened a channel for a vast torrent of water from prehistoric Lake Iroquois and the upper lakes to flow east along the present-day Mohawk Valley. Today, Oneida Lake is a shallow remnant of that time, while the Mohawk River is a mere trickle compared to the deluge that carved the wide valley that later served the region as a transportation corridor from earliest settlement times until the present.

As others have noted, New York was fortunate among the mid-Atlantic and New England colonies to have several places where the prehistoric outflow of the Great Lakes carved gaps in the north–south chain of mountains that make up the Appalachians and New England Uplands. These rocky ridges stretch from northern Georgia all the way to the Canadian border and blocked the westward movement of colonial settlement for a generation or more. The gaps in New York State that were created by the outflow of the glacial precursors of the Great Lakes made the canal possible. And that waterway eventually created the "Empire State." Today, the corridor it follows remains in heavy use. In the Mohawk Valley, the canal, rail lines and the New York Thruway all squeeze through several narrow openings flanked by steep-sided hills.

While dozens of books have been written on the nation-building Erie Canal of the nineteenth century, much less has made it into print regarding the present iteration of the waterway, known for many years as the NYS Barge Canal. In 2000, it was designated a "National Heritage Corridor" by the Federal Parks System, and today it's the New York Canal System run by the Canal Corporation under the NY Power Authority, itself a corporate authority created to further the public interest. The historic designation came about partly because New York's canals have been in continuous operation since 1825, longer than any other man-made transportation system in North America.

The state's canal system was an active commercial waterway until the opening of the St. Lawrence Seaway put most of its specialized fleet of motor ships and barges out of business. However, it remains fully navigable and serves some barge traffic along with recreational boaters and terrestrial towpath tourists. My impression of the current waterway when I first traveled from Oswego to Albany on it with our small yacht twenty-five years ago was that I had entered a time warp. The current version of the canal has operated almost unchanged since 1918. Along many stretches, the stone ruins of the first two canals—old locks, aqueducts, spillways and abandoned factory buildings from the Empire State's industrial age—are prominent features of the landscape. The abandoned structures are powerful reminders of our history and reinforce the sense of an alternate reality associated with canal travel. The riveted ironwork and century-old machinery of guard gates, locks, lift bridges and dams of the present waterway fascinated me as we chugged along. The current waterway is an outstanding example of fully functional analog engineering from the early twentieth century.

The third revision of the canal system follows part of the route of the original Erie plus several natural waterways that served as the region's very first highways in the 1700s. The towns and villages like Seneca Falls, Waterloo and Tonawanda that sprang up along the old Erie in the first decades of the nineteenth century remain viable and rich in history, while the eastern route of the current canal presents striking scenery through the Mohawk Valley that has been compared to Europe's Rhine River.

The working fleet that maintains the canal's century-old infrastructure and machinery along with the shoreline's historic buildings and ruins from the earlier waterways make this a living linear museum. Passage through a lock that uses one-hundred-year-old machinery and equipment, some of which is identical to that of the original Panama Canal's system, is a flashback to the mechanical engineering marvels of 1910. Until recently, the canal system was also a veritable antique tug and workboat refuge. Several steel-hulled tugs that were historic in their own right have served the system since its opening, and several units of the state's canal fleet (aka the "floating plant") have been recognized by the National Landmark Registry. At least for the moment, those vessels have been stabilized, though none are now fully functional.

The canal system faces an uncertain future. There have been controversial proposals to decommission parts of it for various reasons. If this were done, the canal would lose its integrity and fundamental purpose as a navigable passageway for long-distance travel. So it is appropriate to publish a general-interest work on its contributions to our regional history while it remains fully functional. Today you can still time travel along shaded towpath trails, through busy canal-side villages and along serene rivers by boat, kayak, canoe or bicycle. Should you decide to spend a day, a week or a whole summer exploring its natural and man-made history, you'll find a unique coastline like no other in North America.

ACKNOWLEDGEMENTS

I must acknowledge and thank the people who made this book possible, including Andy Derby, Lisa and Dan Wiles, Canal Society's Craig Williams, Will Van Dorp, John Montegue, Jim DeNearing, Rob Goldman, fellow canallers Mary Gwen Todd and Queeno Van Auchen and last but not least my co-captain on the raging canal, Chris Gateley.

Introduction

NEW YORK STATE CANAL SYSTEM

The Erie Canal is part of America's folklore—low bridge everybody down, a life on the raging canal, Clinton's Ditch and much more were all part of my elementary school syllabus in Upstate New York in the 1960s. A mule named Sal was right up there with Paul Bunyan and Pecos Bill in my children's mythology of heroes. We learned how the canal made us the Empire State and how it boosted New York City's fortunes to become one of the leading seaports of the world. Buffalo, at the other end of the canal, was once the eighth-largest city in America thanks to waterborne transportation. Today, nearly 80 percent of Upstate New York's population lives within twenty-five miles of the current canal. But few New Yorkers are even aware of the present-day canal system's existence. It lies mostly unseen behind a screen of trees or reeds as it traverses pastures, farm fields and forests to link towns and villages across the state. Its placid brown waters can be briefly glimpsed at bridge crossings in one of the dozen or so canal-side towns between Albany and Buffalo, but except for anglers, boaters and the hikers and cyclists who enjoy its flat towpath terrain, most New Yorkers today are unaware of its allure.

Yet it still exists. And it still draws travelers from afar. People from California, Canada and elsewhere traverse its historic waters with recreational boats of various sizes and complexities. Today, you can access it with a canoe or kayak or you can travel aboard one of various commercial tour boats on a day trip. In the past, a variety of "expedition"-sized cruise ships offered overnight passages with great food and shoreside lectures and programming on history and hopefully will do so again in a post-COVID era. The mildly adventurous canaller can rent a houseboat from one of several charter companies that provide vessels for weekend or weeklong self-guided excursions.

At least as many if not more canal travelers experience the canal by towpath travel as by boat. They pack their camping gear aboard a bike or plan a route that allows pedaling between bed-and-breakfasts for overnight stops.

Water is amazing stuff. It gives lift as well as life, and the magic of low-speed canal travel aboard a boat lies in buoyancy. Smooth, seemingly effortless, friction-free travel floating down the tranquil path of the first Erie must have seemed little short of miraculous back in the time of dirt roads, ruts, mud and endless potholes. Today's canal traveler gains a very different perspective about the twenty-first-century countryside from that of the motorist speeding down a highway. The canal takes you through a less populated landscape dominated by fields, distant barns, expansive sunlit marshes and darkly forested hillsides. Here a forty-mile journey may take a whole day. Those who journey across New York State by slow boat on this water path sometimes feel as if they have entered a parallel universe, one that runs on a preindustrial time scale. It was then a quieter age, too, and there is time on the canal to exchange a friendly wave or sometimes hold a brief conversation with a fellow traveler or a lock tender in this place of leisurely passing and deliberation. Sometimes, as you float by a village canal-side park, you are watched by the stationary residents. Sometimes they gaze after a passing boat with a hint of longing and curiosity. Where is that boat from? And where is it and its crew bound? Your journey aboard a canalboat is witnessed by others in a way now rare in our modern times of enclosed high-speed travel aboard cars, trains and planes.

Canal travel on a small, slow boat with an open steering station is a lot like backpacking only without the physical effort. You are intimately connected to the natural world around you and self contained with all your needs at hand as you explore byways that are well off the beaten path. There's ample time to savor your surrounding scenery and time also for human contact with other wayfarers. It gets addicting after a while. I, myself a sailor of wide waters in the past, have gotten hooked on it. The canal is a man-made feature embellished and made beautiful by nature in a way I doubt a cement interstate highway ever will be. Aboard a slow, quiet boat, you may hear bird calls from a forested shore. Or you may surprise a deer that has come down to the shore for a drink. A mink might hurry along the shore in pursuit of a fish dinner, or you may meet the gaze of an eagle perched in a tree beside the canal or watch an osprey pluck a fish from its waters. Around each

A fleet of steel-hulled "Lockmaster" boats operates on the canal as bareboat charters. *Author collection.*

bend, something new always seems to await the traveler. Not surprisingly, some people become quite devoted to this mode of travel. Long-distance "loopers," a small subculture of yacht owners who travel with low-speed "trawler" yachts, regularly journey from New York City's salt water to the tides of New Orleans by way of thousands of miles of canals, rivers and lakes in New York and Canada. Some circumnavigate the eastern United States by water, taking a couple of years to do it.

The canal also fascinates modern users as a showcase for sheer human engineering ingenuity. Its century-old locks, rock cuts, dams and its two "overpasses"—at Medina and at the embankment over Irondequoit Creek where it crosses a mile-wide valley—showcase impressive achievements from the analog age. Pedestrians abound on the well-maintained towpath trail between Rochester and Buffalo. Boat traffic these days is light, recreational and often local in nature. For canal travelers willing to take command of a rental boat, a number of boats designed specifically for calm, narrow waters are available for weeklong charters.

Even the most clueless navigator can generally find his or her way down the canal. With the exception of a few lake crossings, its protected waters are suitable for kayakers, canoeists, rowing boats and even stand-up paddle board travel as well as for houseboats and large power cruisers.

Bird sculpture marks the beginning of the footpath along the Cayuga Seneca Canal at Seneca Falls. *Author collection.*

The "Barge Canal," as it was known before acquiring status as a National Heritage Corridor, doesn't see many barges. But the canal is useful, even essential to some businesses that depend on it for water. Its flow-control structures also help regulate the water levels of the two largest Finger Lakes and a number of other rivers and creeks to prevent widespread flooding. Farms draw on it for irrigation, and several manufacturing establishments use its water for cooling and cleaning, while a number of hydropower plants generate electricity from its flow.

So let's shove off for a few hours of modern-day canaling on the fabled waterway that helped build a nation. Follow the route of the nineteenth-century engineering wonder of the world, compared back then to the pyramids of Egypt as a marvel of humanity's ingenuity. Pause to consider historic plaques, overpass graffiti, old stonework and a fair amount of industrial obsolescence as you reflect on the social and cultural contributions made by this waterway. Our society and economy have changed a lot in one hundred years, yet the canal has remained curiously constant. There is nothing else like it in America today.

PART I

HISTORY

1

A BRIEF HISTORY OF NEW YORK'S CANALS

1790 to 2020

Not many people know that fear, along with a desire for profit from trade and real estate, provided a motive for building the first Erie Canal. Before the Revolution, the north shore of Lake Ontario was part of a region known as Upper Canada and was under British control. Memories of the Revolutionary War had not yet dimmed when the first proposal for a canal between Albany and Buffalo was made in 1803. Before the War of 1812, the historic westward route followed by would-be settlers and traders lay well inland of the lake's south shore, then still a frontier bordering on a potentially hostile body of water under British control. The boundary between the United States and Canada was not firmed up until after 1796 when the British finally gave up their military posts at Oswego and the Niagara River. An alternate route for travel and commerce away from the waters of Lake Ontario was desired, and a cross-state canal was seen as essential by some visionaries. Not everyone agreed though. Before the Erie, only three canals in the United States were more than two miles long, and skepticism about canals' practicality abounded.

A second motive for building the canal between Oswego and Syracuse was to eliminate constant flooding from Onondaga Lake and the settlement of Syracuse with its essential and highly profitable salt production pans at its south end. It was believed that deepening the outlet channel of the shallow lake as part of the canal would lower seasonal floodwaters and drain some of the marshlands that were "highly injurious to health," according to a resolution of the state legislature quoted by DeWitt Clinton in his private journal. This would make the land more suitable for settlement even as it also made transport of salt to distant markets far more practical and less costly.

In 1810, Clinton, promoter of nineteenth-century infrastructure and governor-to-be of New York State, surveyed a route for the first canal between Albany and Lakes Ontario and Erie for the Western Inland and Lock Navigation Company. His private journal documenting that survey is online, and it is a fascinating read. Clinton was a sharp-eyed observer of the region's geology and natural resources of potential economic significance, including fisheries, botany and forestry. He also wrote of the early settlement of the proposed route. As an example of his detailed chronicle of the region, he noted: "In dried mullen [*sic*] we discovered young bees in a chrysalis state, deposited there by the old ones, and used as a nest." At another location near present-day Fonda where the highlands crowd close to the Mohawk, he noted the ice age outlet for all the great lakes, observing, "The river must have burst a passage for itself. The opening of the mountains exhibits sublime scenery."

After the War of 1812 was concluded in 1815, construction on the waterway began two years later. It generally followed the course of the Mohawk River and the Wood Creek route as far as Syracuse as a separate line-cut ditch before heading west through Rochester and ultimately to Buffalo. After it was completed in 1825, it was an instant economic success though it was only four feet deep and forty feet wide. The debt that financed it in the form of municipal bonds, said to be the first such notes issued in America, were all retired after just eleven years. After the canal was finished, demand for Syracuse salt increased to a fever pitch. In 1836, a small tax increase on a bushel of salt led to several million dollars of the canal debt being paid off.

Canalboats sized to fit the first locks could carry only about thirty to forty tons of cargo, not much more than that hauled by a modern eighteen-wheeler truck. Yet the modest waterway was a huge improvement over the muddy turnpikes and unpredictable waters of the Mohawk River that had been the highway for settlers and commerce. In the spring, river currents were an obstacle to westbound travel, while often the river's summer flow dwindled to the point where boat travel was impossible. Even under good conditions, there were the two major obstacles in the form of waterfalls at Cohoes and Little Falls that had to be portaged around, so the canal, even with eighty-three time-consuming locks to pass through, was an instant success. The cost to ship goods dropped to a fraction of what Durham boat and ox cart had cost, and its economic effects spread far and wide. Rochester, Utica and

Syracuse were sleepy villages that transformed in a few years into major flour milling and manufacturing centers, respectively, as real estate prices doubled and tripled sometimes in just a few months. And the canal made New York City the Big Apple, the dominant seaport of the East Coast.

THE CANAL "BUBBLE"

The business boom associated with the original Erie led to a veritable frenzy of canal building that was as intense as any modern stock market or real estate bubble. Legislation in 1825 called for surveys of as many as seventeen "lateral" connections to the main Erie, and seven of these were eventually built. They spanned much of the state from the southern tier to the Canadian border. One, the ninety-seven-mile Chenango Canal, connected the Susquehanna River drainage that enters the Chesapeake Bay to the Erie Canal. It ran between the city of Binghamton to the south and Utica on its north end. It was a formidable project that cost over $2 million and required seventy-six locks to climb over seven hundred feet in twenty-three miles. It also required many miles of feeder canals to supply it with water from a system of man-made reservoirs. It operated for just forty years before shutting down. Another notable lateral was the Chemung Canal, which ran between the south end of Seneca Lake over to Elmira. It also connected the Erie via the Seneca Cayuga Canal with the Susquehanna River. It was built over a period of three years and had to surmount a rise of more than four hundred feet between Seneca Lake and Horseheads. Its fifty-one locks increased both building and operating costs considerably, so the locks were built of wood rather than stone. This contributed to the Chemung's short life. One of the major products shipped on it was Pennsylvania coal. The feeder lasted barely forty years. It closed in 1878 after fueling the growth of Elmira, Corning, Montour Falls and Watkins Glen.

An eight-mile-long canal that connected Keuka Lake to Seneca Lake and the rest of the system was approved in 1829. That waterway, the Crooked Lake Canal, followed the Keuka Lake Outlet stream. This short waterway proved exceedingly costly, as it took twenty-seven locks to climb and descend between the two lakes. Though heavily used, its tolls never turned a profit even as the mill owners along the Outlet complained bitterly about the water that was diverted away from their machinery by the canal. In 1872, a drought left just two feet of water in it to float the boats, and in 1877 the

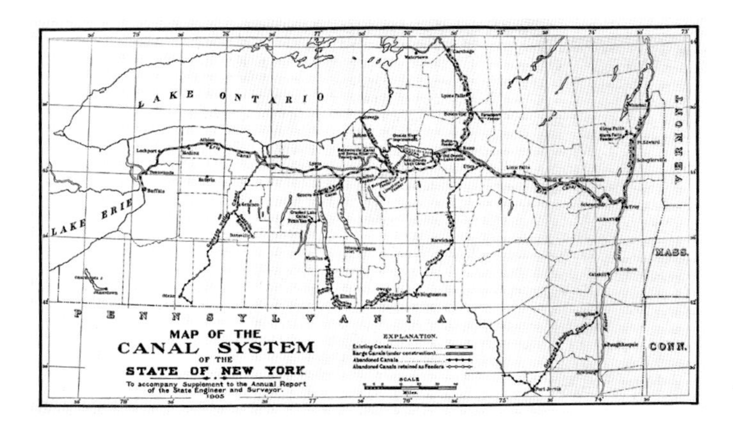

Map of canal system, 1905. *Library of Congress.*

state abandoned the canal. The Penn Yan and New York Railroad laid track along its right of way.

The Black River Canal that climbed Tug Hill was a 35-mile waterway built in part to supply water to the mainline Erie. It lasted longer than the other endeavors, and to this day, parts of it continue to serve as a feeder to the canal system. From Rome, the original canal followed the Mohawk to the junction of the Lansing Kill, where it headed north through a major gorge climbing 693 feet to its summit at Boonville. The channel along this stretch had to be dug from shale, a soft rock that was highly permeable to water and readily formed sinkholes that could swallow a lock or the waters of the canal itself. The towpath excavated along the length of Lansing Kill gorge was constantly prone to washouts and landslides. After reaching Boonville, the canal had to be excavated from limestone, hard granite and gneiss as it descended to Lyons Falls and ultimately to Carthage. It took five years to complete this "engineering marvel" with 109 locks, a total said to be a world record for any canal. The entire original Erie from tidewater at Albany to Lake Erie above the Niagara escarpment rose just 692 feet in its 363-mile length and had 83 locks. The Black River Canal, less than one-tenth as long, climbed just as high. It cost $3.5 million to build, almost half of the cost of the entire Erie. It was the longest-lived of the laterals, lasting until 1922.

All of these early canals and feeders were dug mostly by hand, with much of the hard labor done by recent immigrants, especially Irishmen. The workday lasted from sunup to sundown with short breaks for meals and whiskey. The alcohol kept the men "complacent and working" at the brutal, backbreaking jobs, according to one account. According to a 2015 exhibition based on documents from the National Archives, the

Black River Canal feeder stonework. *Library of Congress.*

consumption of hard liquor reached over seven gallons a year for every drinking-age American in the 1830s. One contractor is said to have replied to a temperance advocate who observed the appalling amount of drunkenness among his workers, "You wouldn't expect them to work on the canal if they were sober, would you?"

The hilly uplands these laterals passed through contributed considerably to challenges of construction and operation. Many streams had to be crossed, and floods were frequent, while the large number of locks and many miles of feeders added heavy costs to operate and maintain the waterway system.

Unlike the various laterals, the main line canal was exceedingly profitable for many years. Work on an enlarged version began shortly after the first canal paid off its debts. The "enlarged" version at seven feet deep and seventy feet wide along with fewer bigger locks could handle boats of 240 tons displacement. It still followed the route of the first canal, running south of Oneida Lake and passing through the city of Syracuse by the south end of Onondaga Lake, but it featured a number of upgrades. These included thirty-two large aqueducts built of stone that carried the canal in a wood-lined trough across creeks and rivers. Numerous stone-lined culverts diverted smaller streams under the canal bed, and a number of spillways and feeders

Remains of aqueduct that once spanned the Seneca River near Savannah, New York. *Author collection.*

with reservoirs and dams had to be built to keep water levels in the canal constant. The reduced number of locks (seventy-two instead of the original eighty-three) made passage faster.

Over the next seventy years, the state continued to upgrade the canal. As competition from the railroads increased after the Civil War, the state reacted by dropping all canal tolls in 1882. This, however, had unfortunate consequences for canal maintenance, and by 1900, the old canal with its animal-powered wooden barges was losing its economic viability to year-round railroad operation. And this was a concern.

Railroad monopolies now increasingly influenced almost every aspect of business. Both industry and the American republic itself seemed to be at the mercy of Big Rail. In the early twentieth century, as one executive put it, you could pay the railroad price or you could walk. There was no alternative for shipping freight in most areas of the country. Without competition, the railroad monopoly could and did engage in abusive pricing as the operators of individual roads cooperated and colluded with one another. The small and exclusive club of rail oligarchs viewed competition as dreadfully wasteful, so they agreed on pricing. In effect, as one contemporary observer stated, they decided to a large degree what businesses would succeed and which would fail. The industry had also received various federal and state government subsidies in the form of land grants, government bonds and favorable legislation. And to add insult to injury, the NY Central Line applied a "differential rate" to freight on its trunk lines across the state that was higher than that applied to trunk lines serving ports in Galveston, Newport News and New Orleans. Not surprisingly, the politicians of New York saw the canal as a vital check on the power of the railroads to set prices, while New York City businessmen saw a modern waterway as essential to the city's continued existence as the major leading seaport on the East Coast. In 1900, shortly after his election as governor, Teddy Roosevelt signed an act directing a survey and cost estimates for a new canal. The 1,020-page report he authorized was objective, unbiased and accurate, and it included a geologic map of the state. This facilitated accurate cost estimates during the bid process for excavation and construction of canal facilities. In the past, contractors had often padded their costs by classifying the material they dug up as containing more rock on paper than it contained in reality. The report's factual basis was essential for the eventual success of the Barge Canal, and canal historian Thomas Grasso credits Roosevelt as being the "father" of today's waterway.

To be truly competitive with the railroads, the canal needed to be bigger with fewer locks and bigger lifts. After the extensive information-gathering process started by Governor Roosevelt followed by a public outreach and a multiyear campaign for a new canal, the people of New York approved a referendum to fund an enlarged modern waterway. They did so partly as a check on the power of the railroads as well as to encourage the growth of industry within the state.

The decision did not come easily, however, because of an unfortunate effort in the 1890s to update the old canal. $9 million of taxpayer money went into a mostly unsuccessful effort marked by extensive corruption and ineptitude. One reason for the financial fiasco was the fact that the state engineer was given just twelve days to create a plan and estimate costs for an enlargement to a depth of nine feet. With no recent good survey to rely on, he made his estimate based on an antiquated data set from 1876. The best guess in 1896 was that the mandated improvements would cost $16 million. (Today's cost in 2022 equivalent dollars would be about $560 million.) The state appropriated a little over half this amount, so the effort was underfunded from the start.

The canal's decline in condition was largely due to a lack of regular funding after the tolls were abolished. The waterway's infrastructure was getting decidedly creaky with frequent disastrous bank breaks; badly needed bridge, culvert and lock repairs; and other issues that caused delays in shipping sometimes for weeks at a time. In more than one instance, cargoes were delayed so badly by breakdowns in canal infrastructure that they were transferred from a stranded canalboat to the railroad. The vital feeders that replenished the canal's water lost from evaporation, locking operations and leaks were often clogged with silt and rubbish. Aquatic weeds choked one stretch of the canal in Wayne County to the point that the mules could barely drag the boats through the vegetation. And some of the locks, built of local sandstone or shale, were crumbling after less than fifty years of service. At the locks, seasonal tenders appointed by the patronage system of the day were sometimes of questionable physical or mental competence. And some tenders were not above squeezing a little private "toll" money out of the boat captains in order to secure timely passage. Cozy relationships between contractors hired to do repairs and state engineers were noted by more than one reform-minded politician and by a number of journalists during this era. In an emergency situation such as a big break in the canal bank and subsequent draining and flooding, a mad scramble of hundreds of laborers, shovel men, carpenters, teamsters, timekeepers, blacksmiths, pile driver operators, steam pump engineers and others ensued to make repairs. Where all the money went during such emergencies was not always recorded. Things were bad enough that the state engineer, Silas Seymour, declared the toll-free canal had been an utter failure, writing that "it must be regarded as a foregone and inevitable conclusion that the canals must go."

During this dark era, more than once where the canal grade was above that of the surrounding land, the embankments were deliberately breached by human action. Motives for such destruction varied, but if a village or town was near a break, the subsequent state-funded emergency action to fill the breach was a financial windfall to local saloons, hotels, livery stables and other businesses. The little logging town of Forestport on the Black River Canal north of Utica suffered a series of failed canal banks over a short period of time.

All of these failures occurred at about the same location, and eventually, canal officials with the help of the Pinkerton Detective agency tracked down and convicted the conspirators who had cost the state's taxpayers tens of thousands of dollars for repairs. (For the complete and fascinating tale, check out the book *The Forestport Breaks* by Michael Doyle.)

Well before any work on the present-day canal began, a number of measures that tightened up the bid process were put in place by the state engineer and his co-workers and advisors to make sure the money went into the canal instead of into the pockets of a "spoilsman" or a sleazy politician. And after years of negotiating, planning and lobbying, when the new canal was given a green light, work on it began almost immediately.

Technology Impact on New Canal

During the life of the nineteenth-century canal, America was swept by tremendous advances in technology, some of which made the new enlarged canal's design possible. A big change was the route. The new canal followed the channels of rivers, streams and lakes for much of its length. This was only possible because of the so-called second Industrial Revolution, as the many advances of the late 1800s are collectively known. One of the drivers of that revolution was Nikola Tesla's AC induction motor, said to be among the top ten greatest inventions of all time. Another major technology development related to Tesla's motor was electrical generation from water power that could be transmitted over long distances via alternating current. The first

Machinery powered by electric motors now opens and closes the lock doors. *Author collection.*

large-scale AC generating plant in the nation went online in 1895 at Niagara Falls, a few miles from the canal's western termination. Whitford's *History of the Barge Canal* mentions one reason for the waterway's enlargement as being the possibilities of cargo transport associated with a rapidly expanding manufacturing complex along the Niagara River between the Falls and Buffalo that occurred after electrification. Here, steel and early chemical industries with ready access to electrical power along with efficient water-based transportation via Erie and other waterways sprang up. Electrical power generated by canal water was also vital to the design, as it was used to open and close the massive doors of the enlarged locks and to operate other lock machinery.

In the early 1900s, new transportation technology was transforming the landscape. Internal combustion engines were rapidly being adopted for use on land and afloat, while steam power was reaching its own peak of efficiency with smaller, lighter high-speed engines. The four-footed hayburner that had towed barges and packets on the canal for nearly a century was no longer needed. Tugs could move strings of barges faster and cheaper. These longer tows influenced the canal's design, as the waterway had to be straighter and wider with periodic turning basins for the tows.

By 1900, the masonry skills that created the enduring and beautiful cut-stone structures of locks and aqueducts of the two earlier canals were becoming obsolete by the use of concrete poured into forms.

Rotary kilns and standardized testing made concrete cheaper and durable enough for wide use on the new canal. Nonetheless, there was controversy and opposition to abandoning the age-old methods of stonework. Much of the anti-concrete sentiment, not surprisingly, was fueled by various labor organizations representing masons, stonecutters and bricklayers. Admittedly, concrete was still relatively new and unproven in 1904 within New York State, and there had been failures of early concrete structures, so a state engineer made an extensive study of concrete structures in other northern areas like Minnesota that had harsh winters. His report determined the new material was as durable as cut stone at about a quarter of the cost. Much of the savings were due to avoiding the costs associated with the massive amount of stone that would have been prepared and transported from quarries to make the various locks and bridges of the new canal. So the ancient trade of the mason and stonecutter took a heavy hit from modern technology along the canal corridor.

Improved engineering and steam-powered construction equipment and techniques along with other advances like the recently developed movable dam and the Tainter gate dam made the modernized canal design and operation possible. No towpath was needed, and the canal could follow natural waterways tracing the routes of the earliest travelers across New York who had used the Mohawk and other rivers. With twentieth-century materials and engineering, flow control structures could now "tame" the rivers and lakes along the route to keep water levels constant (at least most of the time). When the new canal was completed in 1918, shipping moved across the state between Albany and Buffalo through just thirty-six locks, half the number of the old waterway. And the machinery of all the big new locks—including the eleven locks built on the Champlain Canal and the four on the Seneca Cayuga Canal—moved at the touch of a button or switch lever, powered by electricity, with most of the power generated on site from the flow of water through the locks.

2

BUILDING THE BARGE CANAL

The last version of the canal was built in phases and began in 1905. It was an intermittent process that allowed commerce to continue during the enlargement, and it took until 1918 to complete. Despite the availability of steam-powered construction equipment, the task still required an army of unskilled laborers. Rather than Irish, this version of the canal employed labor in the form of recent Italian immigrants along with some Slavs and Hungarians. They were housed in sometimes marginal labor camps under unsanitary conditions while working on the canal and often subject to exploitation by an immigration agent or padrone, who supplied them with food, shelter and other necessities at an inflated price. When workers sent money back to their distant families, the padrone charged them a hefty fee for the service. Though conditions on the canal were far from ideal, some of the immigrant laborers did manage to establish themselves in upstate towns and villages. Eventually, through hard work, they and their descendants managed to claw their way into the American middle class.

The new canal and locks allowed the movement of considerably bigger barges and motor ships with a total displacement of up to two thousand tons. The channel was 12 to 14 feet deep and 120 to 200 feet wide. Much of it followed natural waterways, especially east of the junction of the Oswego Canal, where it followed the Seneca and Oneida Rivers and crossed Oneida Lake. A few miles east of Rome, it joined up with the Mohawk. West of Rochester, the original old line-cut route was followed, but the waterway was enlarged. In some areas such as just west of the landing in Clyde and in

Palmyra, you can see all three versions side by side, the first two being mere marshy depressions now overgrown with vegetation.

While rerouting the canal to use rivers and lakes made good economic sense statewide, the local effect of the new route on towns like Port Byron, Camillus and Jack's Reef that had grown up along the original Erie was little short of disastrous. The canal had been a vital part of the local economic scene, and the new route resulted in the instant loss of hundreds of jobs and businesses. As grocery stores, livery stables and boatyards shut down and stevedores, teamsters, store owners and saloonkeepers lost work, the canal towns became shadows of their former selves. As with former stagecoach towns in the West once the railroad or new highways bypassed them, the now landlocked "ports" and their one-time nineteenth-century waterfront bustle simply vanished. Only their names on roads or old maps recall their existence.

Going with the Flow

Controlling the flow of the natural waterways used by the new Barge Canal was only possible with the engineering advances of the twentieth century. It required far more massive machinery than had been employed on the older, smaller canals. But to this day, that control remains tenuous at times. Water is always on the go, condensing, evaporating, freezing, flowing and sometimes accumulating with stunning rapidity during downpours and floods. Keeping it constant in level and always in place in the canal during droughts and heavy rains requires considerable effort and continuous vigilance. All too readily, water seeks its level, and the tranquil waters of the canal are, in truth, a battleground with entropy as Mother Nature has her own notions of what's proper behavior for a waterway.

The rivers with their highly variable flows that were used by the new canal posed their own set of problems. The Seneca, Oneida, Oswego and Mohawk along with various smaller waters like the Clyde were "canalized" by a series of dams to form a series of twelve-foot-deep navigation pools of constant level. Locks connected these pools wherever water levels had to step up or down according to the terrain. Enlarged feeder canals and waterways were needed to replenish waters lost to locking and evaporation on the new canal, so several larger reservoirs with big fixed dams had to be constructed to supply it.

Oswego's high dam made of concrete to withstand heavy runoff from the river's large watershed. *Author collection.*

The Oswego and Mohawk Rivers posed special challenges to the creators of the new canal. Both rivers have large watersheds. The Oswego's watershed is one of the largest in the state, reaching south into the southern tier counties and taking in nearly all of the Finger Lakes region.

The Mohawk's watershed stretches north far into the Adirondack Mountains, with its headwaters near Boonville, and south into the Catskills. It accepts the volume of the seventy-six-mile-long West Canada Creek, which flows out of the southern foothills of the Adirondacks, and the Schoharie River, which begins near Indian Head Mountain, ninety-three miles south of its confluence with the Mohawk. A heavy rainstorm or snowmelt miles away can cause these tributaries to rise and flood quickly without much warning. In the winter, large blocks of ice form in the shallow river waters, especially in the Mohawk, and in the spring, these chunks create ice dams that back up the water for miles. Once the ice dam is broken up, the rushing water and ice chunks can destroy anything they come into contact with.

Large fixed dams for the canal were not feasible on the Mohawk both because of ice and because raising the level of the river to create a reservoir would put large areas of the adjacent developed farmland and villages along the Mohawk under water. This problem led to the installation of several

Movable dam at Lock 10 on the Mohawk River. *Author collection.*

movable dams. The dams along the Mohawk are among the last of their kind and are today considered engineering landmarks.

The movable bridge dam is a structure of steel gates and supporting framework that can be raised or lowered to change the level of the river. This creates a temporary pool of water for navigation during the ice-free season without creating a huge reservoir that would flood the surrounding area. The first movable dams were built on European canals around 1834, and the earliest versions consisted of a series of iron frames fixed to a foundation across the river bottom. When not in use, the frames lay on the bottom of the river. When needed, men would lift the frame upright and lay a wood walkway on the top of the frame. Then the next frame was lifted and the walkway installed. Once complete, the entire affair looked like a road trestle, and the structure was sometimes called a trestle dam. The river water was then blocked by installing long, narrow boards set on end against the frame. By installing board after board, a dam was slowly built up and the water impounded. The height of the pool depended on the height of the trestle and boards. If the pool backed up behind it had to be regulated, some of the dam boards could be removed. And if the natural condition of the river allowed navigation without dams, all the boards could be pulled and the frames returned to the river bottom.

New York State hired David A. Watt to create a modernized version based on a dam built on the Moldau River in the Czech Republic. Watt realized that on the Mohawk, unpredictable seasonal flooding and ice jams were a major issue. Ice would quickly damage or destroy frames lowered to the bottom of the river during the spring freshets, so Watt opted for installing a dam with its frames suspended from an overhead bridge so they could be lifted clear of the water. This kept them safe from the ice and flood debris and made them easier to maintain. It also allowed the river to run free, and spring's high water would flush out the silt and mud that accumulated each navigation season behind the dam. This reduced the need for and expense of periodic dredging.

Watt built eight of these innovative structures along the Mohawk between Schenectady and Fort Plain, double the total of such dams then in use around the world. They featured multi-span truss bridges to cross the river. The locations of the dams placed them squarely in the sight of the public, as the New York Central Railroad and the main highways ran along the river, so their successes or failures would be on view for all to behold. As it turned out, a century later the dams are still in service, though they have been rebuilt and modernized. Construction began in 1906 and finished up by 1913. For the most part, the structures are alike, although adapted to their specific location. All the dams east of the confluence of the Schoharie have three bridge spans, as the river was naturally wider with the additional flows from the Schoharie. All the dams west of the Schoharie have two spans.

The level control frames are lowered into the river channel by means of pivoting arms so that their bottom edge rests against a concrete sill that spans the river's bottom. With the frames in place, two large steel plates are lowered or raised to regulate the water flow. Each dam has a lower and upper plate that moves independently of the other rather like a sash window to control the volume of water passing downstream. During normal river flows, all the lower plates are left in place. If conditions demand a lowering of the pool, the upper gates are raised to increase flow through the dam. Control of the river must be coordinated with lock tenders and crews who operate the movable dams downstream. If too much water were to be released, it could cause flooding. In the off-season and during the spring freshets, the gates are lifted completely clear of the water, allowing the river to flow naturally.

Monitoring the flow of the river is a constant task for the lock tenders here. Is the rain simply a brief shower causing a small localized rise in the water, or is a major flood in the making? The lock tenders who also look after a dam keep a close watch on precipitation forecasts and river level data, as waiting too long to raise their gates could be disastrous. The movable dams, cutting-edge innovations of hydraulic engineering, were soon tested.

In March 1913, a late winter storm that brought widespread floods to the Midwest moved east to hit New York. Unfortunately, the dam gates at Locks 12 and 13 had been lowered to allow the dredging contractors to get an early start on deepening the channel downstream for the season. As the floodwaters rushed through the valley and various tributary streams, they collected lumber from Barge Canal contractors and other sources. This material, along with uprooted trees and other debris, quickly caught and jammed up the frames and gates of the lowered dams. The water behind the dam then backed up to a greater depth than the structure was designed for. Disaster ensued as the pressure caused the plates and frames to twist and bend, chains to break and supporting members to buckle. And making things worse, once the debris and logs built up against the gates, there was so much pressure that the operators couldn't raise the frames.

After this first flood, all the movable dams were reinforced. The steam-powered winches that had raised and lowered the gates were rebuilt to run on electricity, allowing for a quick start up with the flip of a switch in an emergency. The lifting points for the gates were modified, and the reinforcement saved the dams from future flooding damage. Until 2011. That's when Hurricane Irene and Tropical Storm Lee brought thirteen inches of rain to the valley and caused over $50 million in damage. The dams held, but at a severe cost (see chapter 7). Because of the difficulty of operation during flood conditions, the movable dam has fallen out of favor. It has largely been replaced by other types of controllable dams like the Tainter gate or spillway dams. The canal's movable dams are among the last of their kind. It remains to be seen how well they will cope with the increased extremes of weather associated with climate change.

AQUEDUCTS AND EMBANKMENTS

The eastern part of the canal largely follows rivers, but the western section remains largely a line cut with a long stretch of level water making up a pool. Here, the traveler encounters two examples of one of the more spectacular oddities of canal engineering, the embankment, a sort of canal overpass across a valley or ravine. As you head west, the first embankment crosses

Irondequoit embankment in 1921. *NYS Archives.*

the wide valley created by Irondequoit Creek's ice age precursor just east of Rochester. The Great Embankment is a gigantic man-made ridge of fill supporting a trough of concrete that holds the canal's water. The original Erie Canal followed several natural ridges that were linked together by the embankment to get across the valley. One man-made ridge was 1,320 feet long, the other 241 feet. These massive piles of dirt had to be excavated by hand and moved by horse-drawn wagon in 1822, as steam shovels and diesel-powered excavators lay far in the future.

Today, the seventy-foot-high embankment carries the canal above the rooftops and trees of the land below for over a mile, and it's said that on a clear day you can look north up the valley to see Lake Ontario. Thousands of tons of soil and gravel buttress the canal trough. The two sides of the embankment are twenty-two feet thick at the top and slope down to the valley floor on a one-to-three angle (one foot down for every three out). Obviously, a leak in the canal bed here in the heavily developed suburban environment that fills the valley today would be a disaster. This did indeed happen to the predecessor canal—not once, but several times. Considerable effort has gone into preventing a blowout like the one that took place in April 1871, and to this day, bank walkers keep an eye on the canal's integrity here and wherever else it runs above grade. The present-day canal also has a guard gate at each end of the Irondequoit embankment to limit drainage and flood damage should the trough spring a leak.

The famous 1871 blowout of the previous canal here was attributed to a busy muskrat whose burrow into the bank turned into a leak. Eventually, over two hundred feet of the canal bank washed away. The flood that followed this collapse abruptly grounded dozens of boats for many miles as it drained the canal, even as the water "with mud and disease rushed over the canal's banks with brute force," according to a contemporary news account.

Hundreds of workers were called in to make emergency repairs. Conditions early in the season when spring snows and mud prevailed were grim, and the hasty assemblage of men soon grew "aggravated" by a combination of bad weather, ample whiskey and tough working conditions. When offered a pay raise, it was violently rejected as the workers became "extremely agitated" and rioted, going so far as to shove a hapless horse into the canal bed.

In 1911 and again in 1912, disastrous breaks drained a large section of the canal. The second failure of the canal occurred just a few weeks after the embankment trough had been widened and straightened as part of the new Barge Canal enlargement. It took six years to fully redesign and repair the damage, and today's embankment is considerably beefier than the nineteenth-century versions. The 1912 structure has a concrete-lined trough whose sides are heavily backfilled with earth. It withstood the test of time for sixty years. But in October 1974, a crew excavating for a sewer line managed to break through the concrete bottom of the canal, causing a catastrophic deluge. Eighteen homes were badly damaged by the sudden release of water, and at least one house was completely swept away. Power and phones were taken out, and gas lines broke as the water surged forth, eventually entering Irondequoit Creek's channel. According to an article from the *New York Times*, about fifteen square miles of the suburban area was flooded with four feet or more of canal water and a number of the residents compared the catastrophe to a popular disaster movie involving a capsized ocean liner, *The Poseidon Adventure*, that had just shown on television. There were no fatalities, but it took many weeks to repair the breach and get the canal back in commission.

A second canal overpass exists just east of the small city of Medina and is the only aqueduct on the current canal. Though shorter than the Irondequoit Bay embankment, it is no less impressive. Here the canal crosses Oak Orchard Creek gorge by means of a massive concrete arch and trough, and towpath pedestrians can look directly down on a 35-foot waterfall. Originally, the canal designers proposed to cross the gorge with a straight section of canal supported by a concrete arch aqueduct spanning 285 feet. Such a design would have been cutting-edge and unprecedented. The structure would

Medina aqueduct looking east showing Oak Orchard Creek gorge to the left and canal on extreme right. *Author collection.*

Here the second version of the canal crosses the Mohawk on an aqueduct near Rexford a few years before it was closed down. *NYS Archives.*

have supported 12,400 tons of water and would have weighed 46,000 tons. Despite a series of tests showing the strength of concrete, the design for a free-standing aqueduct had to be abandoned because the red sandstone that made Medina's architecture famous was suspected of being too weak to support the arch ends. Instead, the canal is supported by a more conservative solid structure made of specially formulated high-strength concrete. It had to be poured in layers within the wooden forms, and the whole job had to completed in a few months during the winter when the waterway was closed to navigation. It's a weird sensation today to stand beside the placid stillness of the canal filling the trough and look over the railing into the deep gorge at a cataract of foaming white water falling into Oak Orchard Creek far below. Depending on the season, you may also look down on anglers in their waders or fishing boats in the creek.

After years of work, the new Barge Canal was completed and opened in the spring of 1918. The old channel was left drained for the last season. A way of life in the little villages along its eastern portion was over. No more would the mules and horses and their footsore drivers tread the towpath, and the small family-operated wooden canalboats and barges were quickly abandoned as wartime shipping under federal control took over the canal.

3

CANAL COMMERCE

1917 to Present Day

In August 1917, America entered World War I. The nation was, however, ill prepared for the vast production of goods, food and military equipment needed to fight in Europe. Everything had to move by rail from farms and factories to the East Coast. And the railroad tracks and equipment were in poor condition after years of competition, labor unrest and cost-cutting. There was no central management to meet wartime needs, so in short order, an epic logistics snarl ensued. Rail car shortages were acute in the interior, congestion in Chicago's rail yards was all but unworkable and vast numbers of boxcars sat empty on sidings on the East Coast with no freight to carry west. By fall, a shortage of more than 150,000 cars left the Northeast region with an acute lack of food and coal for winter heat stockpiles. The ships waiting to take food and arms to Europe had no coal for their bunkers to get their cargoes across the sea. It was a supply chain snarl far worse than the height of the COVID-related delays of 2020–21. William McAdoo, appointed by President Wilson to run the nation's railroads, described it as a "run down chaotic confused mess." The federal government's decision to nationalize the rail system for the war effort was the first time an entire industry had been taken over.

As part of the effort to relieve congestion, the Railroad Administration assumed control over the movement of cargo on the newly completed State Barge Canal as well as on some of the country's western waterways. Unfortunately, few boats designed for operation on the just completed Barge Canal had been built. Most of the canal fleet from the previous waterway consisted of small wooden barges owned by family-operated businesses, while suitable tugs and steamers available for towing were scarce. Steel for building new barges was in short supply in a wartime economy, and existing tugs were busy with work on the East Coast. This led to an interesting but less than successful experiment in ferro-cement canalboats. The Emergency Fleet Corporation was established to build a number of concrete barges. In the spring of 1918, the Concrete Ship Section of the Emergency Fleet Corporation designed a concrete barge for specific use on the Barge Canal. The barge was 150 feet long, 21 feet wide and had a depth of 12 feet. The cargo carrying capacity was to be 500 tons with a draft of 10 feet. The barge was divided into three sections and designed along the lines of the old enlarged Erie canalboats with a forward 15-foot section for storage and crew quarters, a cargo area of 115 feet and a well-appointed captain's quarters of 20 feet back aft. The captain's quarters had a galley, living room, bedroom and wardrobe. Twenty-one barges were built between 1918 and 1919 in various yards. As ships go, the barges were less than totally successful; however, the results of this experiment did advance the engineering of concrete structures considerably.

Several New York shipyards, including one in Tonawanda, one in Ithaca and one in Fort Edwards, built the barges. The hulls were shaped by forms made of wood and were three inches thick except at the bow and stern, where up to four and a half inches of concrete were used. Steel rebar reinforcement was laced together to follow the shape of the outer form. An inner form then created the space to be filled with concrete, and the hulls were created with a single fifty-hour pour. The bow and stern of the barge had some curves, so the skills of the men were tested as they bent the steel reinforcing bars to fit into the three-to-four-inch space available between the forms. The end result was a nicely rounded hull of pleasing lines that was, unfortunately, quite brittle. Grounding or minor collisions with docks often resulted in damage, and a number of the barges soon ended their lives as parts of lengthened lock approach walls on the Mohawk after the war. Their concrete hulls were grounded in position and filled with stones and soil. As the years went by, the concrete flaked and cracked and fell off, and the resulting exposed rebar was notably hard on the topsides of any small recreational traffic that tied alongside. So the remains of the barge hulls were buried and new approach walls were built.

While wartime freight volumes were relatively low, the canal was used to transport a number of military vessels, including minesweepers and sub chasers, from Great Lakes shipyards to the coast. And wartime demand caused a considerable amount of flour, kerosene, gasoline and knit goods to be shipped on the canal.

After the war, cargo volumes increased slowly over the next decade as wooden steamers and barges gave way to steel hulls and larger more capital-intensive ships. In the 1920s, a fleet of barges and several motor ships operated by Standard Oil went into service making deliveries of fuel oil, asphalt and other products to tank farms located at cities and villages on all four of the enlarged canals. Other bulk cargo, including pulpwood and grain, also began to pass along the canal in steel barges. A few barge loads of perishables made their way to New York City. There were even a few loads of live eels shipped from Quebec to New York by way of Lake Ontario and the Oswego Canal in 1920.

When the expanded waterway first opened, wartime shortages of steel restricted any building of new barges. A few new wooden barges of six-hundred-ton capacity were built locally, and several old fish tugs from Lake Erie were converted for towing. A few old wooden towboats that had worked on the previous canal also still steamed along. Richard Garrity in his memoir of working on the canal, *Canal Boatman*, recalls that several of the wooden grain barges were involved in accidents when approaching a lock. If they sank, the wet grain usually swelled enough to badly damage the wooden hulls, causing a total loss to the owner. He wrote that the sinkings were due in part to the use of long hawsers by towboats and tugs. This, along with old-fashioned steering still in use aboard the barges, caused a number of the accidents. While a long hawser had been serviceable with mules pulling smaller, slower barges, it was difficult to control a barge towed at higher speeds at the end of a three-hundred-foot-long line, especially if the barge was equipped with simple manual steering gear. Sometimes the barges even rammed the shore when being towed around a canal bend. The state soon mandated towing bridles and shorter hawsers. Within a few years, tugs began to push rather than tow their barges for better control. Today, the articulated tug barge (ATB) setup, whereby the tug fits its bow into a notch to push the barge, is widely used on coastal waters and the Great Lakes as well as on the canal.

Soon after the transition to peacetime and the Roaring Twenties, several types of specialized vessels designed for moving cargo on both the canal and open water were placed in service. Among these "seaworthy" hybrid canal/coastal ships were the ships of the ILI fleet, designed for carrying midwestern grain from the head of the Great Lakes to saltwater ports along the coast. They were squat, low-slung versions of traditional "laker" with

A canal motorship of the Interwaterways fleet capable of open water or canal operation sister to the *Peckinpaugh*. *Will VanDorp.*

a wheelhouse forward and crew quarters aft and were sized to carry 1,500 tons of freight on a ten-foot draft through the canal locks. The first of these was a little ship that worked the canal for many years under various owners, concluding its service as the *Day Peckinpaugh* (see chapter 5 for more). Before long, motor ships became a common sight on the canal. They were designed to operate on the Great Lakes or coastal salt water, even as they were able to slip under low bridges and through the narrow waters of the canal.

Among these specialized canal carriers were the motor ships of the Ford Fleet. Henry Ford first became acquainted with the canal during a cruise aboard his yacht the *Greyhound*. He had several plants producing cars on the coast and saw the canal as a logical route to transport parts and sub assemblies to the Ford plant in Troy and to other destinations. Ford Motor Company ordered two three-hundred-foot cargo ships specifically designed to fit the locks and pass under low bridges. They were powered by oil-fired steam turbines, an innovation for the time. Steam turbines were more compact, fuel efficient and powerful than contemporary internal combustion power of the day. The keels for the *Chester* and the *Edgewater* were laid at the Great Lakes Engineering Works of River Rouge, Michigan, in 1931.The motor ships had retractable pilothouses that lowered into a well in the ship's hull to allow the craft to pass under bridges on the canal. Other topside equipment, including exhaust stacks, masts and flag poles, could be lowered. All ship controls operated directly from the pilothouse, a novelty at a time when most vessels still used a system of bells for the helmsman and captain to signal the engine room. Each motor ship had a crew of twenty-two men who worked on a 24/7 schedule. The ships were powered by two eight-hundred-horsepower Westinghouse steam turbine engines, fired by fuel oil. They cruised at about eleven knots and had a capacity of 2,800 net tons loaded through nine hatches. Dual rudders and direct reversing engines made the motor ships more maneuverable in the confined waters of the canal.

The two ships proved profitable, and Ford expanded his fleet with two more in 1937. By this time, the company was also running quite a fleet of lakers that carried iron ore to Ford steel mills on the Rouge River. Occasionally, the motor ships would be chartered to haul commodities for companies other than Ford, when the car business slowed down.

The new ships had a capacity of three thousand net tons, the intended limit of Barge Canal locks. The retractable pilothouses were placed farther aft on the new vessels, rather than in the far forward position on the original craft, and the total horsepower was reduced from 1,600 to 1,200, with an eye toward economy. The vessels were powered by newly designed Cooper-Bessemer diesel engines. These ships were the first freighters on the Great Lakes built with completely welded hulls. (In a shipyard, arc-welded hulls went together far faster than riveted hulls with substantially lower labor costs. This technology made possible the launch of nearly three thousand Liberty ships between 1941 and the end of World War II.)

Ford's canal fleet was drafted by the U.S. Navy for wartime service on salt water, as were a number of other canal motor ships. The vessels, modified by the navy with fixed pilothouses, never returned to the canal, and at least one ship was lost to enemy action.

GOLDEN GRAIN

Shortly after the Barge Canal opened, grain—mainly in the form of wheat, along with corn, barley and rye—became a major cargo. More than one million tons were shipped in 1931. In New York City, Albany and Oswego, the state built large grain elevators for storage of canal-shipped grain. The Oswego elevator, completed in 1924, had a capacity of one million bushels in twenty-seven concrete silos ninety-four feet high and twenty feet in diameter. One side took the grain in from ships; the other fed it into canal barges. The material was weighed in two-thousand-bushel batches.

The state expected that lakers loaded with grain that came to Oswego would be able to carry coal from the port's several rail trestles back west.

Oswego grain ship loading at the elevator around 1932. *Richard Palmer.*

Montreal, the other major destination for grain ships, had no such return cargo. However, when the Welland Canal was enlarged, it became more efficient to store the grain from a laker at Montreal, where it could be loaded directly into a seagoing freighter, than to load it from Oswego's elevator onto a barge that would then move it to another elevator in New York City for export. Also, as the Depression took hold worldwide in the 1930s, many European countries put tariffs on imported grain to protect prices for their own farmers. Grain tonnage on the canal dropped quickly, and the elevator's use declined. When the St. Lawrence Seaway opened in 1959, grain bypassed Oswego and the canal. The elevator was used in later years by Genesee brewery for storing barley that was then trucked to the company's malthouse in Sodus. For a few years, surplus corn was also stored in it. The elevator was demolished in 1999, a task that took some time and a good deal of effort, according to at least one memory.

Today, the H. Lee White Marine Museum is located in the one-time office building for the elevator. I recall a few of the lead scale weights were used for trimming ballast during the construction of a steel sailing boat built by the Oswego Maritime Foundation in the 1990s. A few barge loads of corn have

Oswego Harbor from the air in the 1930s. *Richard Palmer.*

been shipped in recent years from Oswego to the ethanol plant in Fulton using the canal. Nowadays, the Port Authority uses a recently built steel storage silo on the east side of the river to store up to 720,000 bushels of grain. The new facility is said to be able to store 15 percent of the soybean crop produced by all of New York State and, the plan goes, will be used for exports by way of the St. Lawrence Seaway.

Another major cargo by volume on the canal between the 1930s and the 1960s was oil in various forms. During the 1940s and '50s, fuel oil, primarily used for heating, made up one of the biggest volumes of cargo on the canal.

In 1948, forty-one counties in the state depended on the canal for their heating oil deliveries. Tank barges carrying up to nineteen thousand barrels of oil served bulk storage facilities at Rochester, Utica, Syracuse and Buffalo. These were of the articulated type, with a notch in the back for the bow of the pusher tug. This allowed more space for cargo, as it shortened the length of tug and barge in the canal lock and also improved overall maneuvering. Various forms of refined product and asphalt were carried by barges. Barges used the Champlain Canal to carry jet fuel to the air base at Plattsburg that closed in 1995, and refined product also went to Rome's air base by way of barge.

Other bulk cargoes moved by barge included iron ore from Port Henry on Lake Champlain, molasses used in animal feed, phosphate rock for

Oil barge locking through during World War II. *Library of Congress.*

fertilizer, various forms of steel as scrap or product, sand, gravel and stone. Soda ash from the Solvay plant in Syracuse went by canal to New Jersey, where it was used in glassmaking and in the production of various chemicals and detergents. Another specialty cargo was that of Canadian newsprint for the New York City market. Barges built for that trade were smaller so as to pass through the locks of the Chambly Canal that connects Lake Champlain to the St. Lawrence. Additional newsprint from Canadian mills came across Lake Ontario to the Oswego Canal. The Barge Canal's peak years for volume of shipping were from the 1930s through the 1950s. After 1951, volumes began to decline as trucks and pipelines reduced the need for shipping on the canal.

Canal Commerce Dwindles

In 1935, with some federal funding, the state began to deepen the Oswego Canal and the eastern section of the main canal from the Three Rivers junction to Albany. Bridge vertical clearances were also increased to about twenty feet to accommodate more traffic to and from Lake Ontario. West of the Three Rivers, bridge clearances remain fourteen feet, eight inches, at normal pool level. The improvement took some years to complete. Cargo tonnage peaked at just over 5 million in 1951. Then volumes began a slow, steady decline. When the St. Lawrence Seaway opened in 1959, it had an immediate effect on canal traffic. Large ships could now pass directly from the upper Great Lakes to Montreal and overseas. Also around this time, the federally subsidized Interstate Highway System began to fuel an increase in long- and short-haul trucking. This, along with increased oil pipeline capacity, sharply reduced the canal's business in the 1960s. Trucks and pipelines could operate year-round, reducing the need for oil tank storage farms. In 1950, 416,875 tons of oil were moved on the canal to Rochester. After a pipeline extension in 1954, the amount shipped on the canal to the city was just 33,159 tons per year.

Road building has a long history of federal subsidies in the United States. During the Great Depression, various WPA projects improved highways, and as early as 1944, federally designated interstate roads and funding for these all-weather limited-access highways laid the groundwork for the long-haul trucking industry (as well as for much of the sprawl of suburbia that continues to chew up agricultural lands and forests in the eastern United States today).

Another factor causing the decline of bulk cargo transport was the gradual deindustrialization of the Great Lakes region, including New York State. Factories moved first to the nonunion Sunbelt South and then ultimately to overseas locations. Manufacturing in general peaked in the early 1950s in the Great Lakes region. After that, all-weather paved highways and the St. Lawrence Seaway simply outcompeted slower canal transport. By 1990, the commercial era of the canal was all but over. Today, huge warehouses dot the inland landscape near highway exchanges, and the few remaining canal terminals have been repurposed to serve as museums and canal workshops or for other public access.

Just as the federal government subsidized railroad transport in the nineteenth century with favorable financing deals and vast giveaways of land for rights-of-way, some of which the railroads later sold for profit, the rise of internal combustion engine technology prompted government subsidies for the Interstate Highway System in the 1950s. The industrial manufacturing powerhouse of the Great Lakes, nourished in large part by the proximity to waterways for economical movement of ore, coal and steel, gave rise to the Steel Belt. But as the Depression-era Tennessee Valley Authority's low-cost energy subsidized by the federal government attracted manufacturing to the South, Rust Belt factories closed, and today much of the old industry that once fronted the canal has vanished.

As the canal's value to commercial shippers and the associated tolls declined, efforts began to shift responsibility for its maintenance to the federal government. Talk of making it a ship canal by enlarging and straightening it and raising bridges quickly stalled when the costs were totaled, and then there was the small matter of the consequences to the state's lakes and rivers that might be drained to supply giant locks needed to accommodate the ships. Some New Yorkers worried that if the feds controlled the canal, it would no longer be available for power generation or flood control. So the idea was dropped.

Canal as a Recreational Corridor

In 1992, the state's Department of Transportation gave up control over the canal and turned it over to the Thruway Authority. An increasing emphasis on the canal as a corridor for recreation and tourism was put in place and remains in force today. In 2017, the canal came under the jurisdiction of the New York Power Authority, which operates it as of this writing. It was given a historic corridor designation under the U.S. Parks System in 2000. Today, a number of local historical societies and museums have exhibits and programs detailing the canal's past and present, and the Canal Society of New York State, founded in 1956, preserves and researches its past history while also supporting ongoing revitalization efforts and recreational access to the current waterway (see chapter 9 for more on the work of these groups).

Today, most traffic on the canal, like this small sailboat at Seneca Falls dock, is recreational. *Author collection.*

PART II

ANALOG ENGINEERING ON THE NEW ERIE

4

LOCKING THROUGH LEVEL CHANGES

Locks are the heart of a canal that traverses changes in elevation. The concept of using locks to control water levels has been around for a long time. The pound lock that contains a chamber of water and allows it to rise or fall is said to have been first used on China's Grand Canal around AD 983. Early pound lock chambers were closed off by gates that moved vertically, guillotine fashion. Nearly all of the locks on the various versions of the Erie Canal have had miter gates, said to have been invented by Leonardo da Vinci in 1497. It's a simple, rugged design and is all but foolproof if the lock components are maintained, and to this day it continues to be widely used on canals worldwide. The miter lock consists of two gates, each on a pivot or hinge, which swing open in unison. When closed, they press up against each other at a forty-five-degree angle so that the pressure of the water against them forces them together.

More water depth and associated higher pressure simply make them close more tightly. But when the lock fills or empties to equalize levels on either side, the gates open with ease. On river sections of the canal, they are paired with a dam. On the land-cut sections they raise and lower the water according to the area's terrain. Each section of canal between the locks is known as a level. Picture each level as its own small lake, and the lock takes you from one to another.

Today's canal locks have massive gates made of steel to close off the concrete- and steel-lined chambers. The elevation change, or lift, ranges from six to forty feet. The side walls of the locks have extra height to allow

Lock 2 at Waterford. *Author collection.*

for higher water in the levels, and all the locks have at their lowest level at least twelve feet of water in them. The total depth from top to floor of the Little Falls lock chamber, the highest lift on the system, is up to eighty feet when the lock is drained for maintenance. The concrete of the side walls of this lock is up to thirty-four feet thick. Within each side wall is a culvert that fills and empties the lock. On the highest lift locks, these tunnels are big enough to walk through with dimensions of seven by nine feet. They are connected to ports in the bottom of the lock that fill and drain the chamber. The flow of water into the culverts is controlled by valves that are raised and lowered by cables driven by electric motors. Originally, since many of the locks were in rural areas that had not yet been electrified, they used hydroelectric turbines driven by their operation to power their machinery. Today, some of them still have their powerhouses, but the power now comes from the grid.

The steel doors weigh about one ton per vertical foot, so each gate of a forty-foot lift lock weighs forty tons. They're driven by either electric motors that power a rack and pinion system or, in more modernized locks, by hydraulic pistons. The quoin post of each gate swings on a cast-iron pivot set in concrete at the bottom. This pivot point has to be periodically renewed, usually during the ten-year maintenance cycle for the locks. White oak edges seal the lock gates, and they take about a minute to open or close.

Oswego Lock 8 seen from the bridge at the entrance to the Oswego Canal. *Author collection.*

Lock 2 of the Cayuga Seneca Canal with eastbound boats. *Author collection.*

Oswego hydropower generator with dam and lock in foreground. *Author collection.*

When I was around seven, my family went for a drive on a midsummer evening to Lock 30 near Macedon. I don't recall the motive or who initiated the excursion. Perhaps it was my dad, who had a long-standing interest in the realm of maritime affairs. He had sailed a few times on his brother's sloop and once began a brief effort to restore a small wooden sailboat before other family needs like digging a new leach bed by hand overwhelmed the project. Or possibly, I sparked the notion of seeing a canal tug in action after rereading one of my favorite stories, *Lil Toot the Tug*. At any rate, the canal was the closest place for us to observe waterborne commerce and tugboats. So, like many a family before and since, we sought diversion by barge watching at a lock.

There is something a bit magical about seeing a boat rise up or sink down into the lock chamber. It appears effortless and largely silent to the observer. And adding to the interest is the chance of seeing something special down in the water or attached to the lock walls. I once saw a tiny sprig of vegetation growing from a crack near the lock's high waterline with a one-inch frog hanging on to it. I wondered how many empty and fill cycles he had gone through. I've also watched more than one goose or duck family "lock

Oswego lock powerhouse showing remains of nineteenth-century hydraulic canal stonework in foreground. *Author collection.*

through" along with a small boat. On my long-ago family excursion, I was delighted by the passage of a black-and-red tug with an oil barge ahead. At close quarters, I was able to take in the patterned curtains at the cabin ports, the incomprehensible jumble of pipes and valves on the barge's deck ahead and the quiet competence of the crew at work handling lines and their tug as the great doors of the lock slowly swung open. To this day, tugs still intrigue me. They were the canal mules of my childhood.

There is one lock located where the canal crosses the Genesee River south of Rochester that is different from the miter gate structures. Here, the river level can fluctuate either above or below the water level of the canal depending on its flow, so a guard lock is needed at the river crossing. Because mitered lock doors work only if there is water pushing them shut on the upstream side, this lock uses a big steel plate that lowers much like those of the guard gates or movable dams to close off the river when needed. Most of the time, the lock isn't needed, and the doors are left open.

Locks are operated by a chief lock operator, a canal structure operator and a seasonal lock operator. In the past, during the canal's busiest years of commercial activity, barges and tugs ran at night, and the tenders worked

Guard lock on the Genesee River has a vertical gate rather than miter style to compensate for larger changes in river level. *Author collection.*

twenty-four-hour shifts. Today their hours are reduced, as recreational traffic rarely runs after dark. The typical schedule for a chief is to arrive at 6:30 a.m. and then do a walkabout inspection before the day begins. He or she checks to see if all is in order and that no creatures on nocturnal wanderings have fallen into the lock. (One lock tender told me he found a hapless lynx floating drowned in his lock on the eastern Mohawk.) Nowadays, most of the locks open on demand beginning at 7:00 a.m. after being contacted by VHF radio channel 13 or by cellphone. During peak summer season, the busier locks work until 10:00 p.m. When not actively operating their lock, the crew does routine maintenance, greasing, painting and chipping rust. Some of the locks have fairly big grassy areas around them to mow, and at least in the past, a number of lock tenders also took pride in tending a tidy flower bed or two. Most of the traffic these days consists of small private yachts and fishing boats with the occasional kayak or canoe. On the sections of canal that follow rivers, the lock crew also is responsible for looking after the associated dam, and during heavy rains, they may be called out at any time to operate dam machinery.

One tender told me that his days from mid-May to mid-October were generally routine, but now and then, like a basketball player on the bench, suddenly the coach would send you into the game. Then anything could happen. Events could range from a sudden downpour and rising river level, requiring immediate attention, to the dam to human-relations management. Angry waterfront cottage owners have been known to accost a lock tender demanding to know why the canal is flooding their yards. A lock tender told me of an incident where a big powerboat threw a huge wake up as he sped down the canal, tossing a camp's canoe up into a waterfront flower bed. The irate landowner chased the boat down to the next lock, and the chief had to intervene to make peace. The lock tender's job sometimes requires the skills of a marriage counselor, engineer, policeman and peacemaker. Thankfully, serious incidents are rare, but now and then they happen.

Inept boaters on their first canal trip can get into trouble in the lock if they don't keep their boat tethered to the lock wall. As you enter with your boat, you'll see a series of heavy ropes or sometimes cables hanging down into the water at regular intervals along the lock wall. The idea is to use a boat hook or to grab a rope and hang on to keep the boat close to the wall as it changes elevation.

Experienced canal boaters keep a pair of gloves handy, as the submerged length of rope gets pretty slimy as the season progresses. Sometimes a novice

Tainter gates of the dam beside the Cayuga Seneca Lock 1. Lock tenders are also responsible for operation of the dams. *Author collection.*

Above: Author near the end of her rope in Lock 17, the biggest elevation on the system. *Chris Gateley.*

Opposite: Drained lock at Lockport—note the holes at bottom of wall for water passage in and out of lock. *Library of Congress.*

may make the heavy hand line rope fast to his vessel, only to find as his boat descends it's now too tight to undo. Lock operators are trained to watch for this. The lock operators are also good at getting little boats through with minimal turbulence, but it takes time to stop the level change once the water begins flowing. They typically keep a hatchet handy so they can cut a boat loose if need be.

By the end of the summer, the routine begins to wear thin for some of the workers, but then comes cooler weather and with it off-season maintenance with its own rhythms and challenges. At this time, the heavy maintenance gets done at the lock or at one of the maintenance shops along the canal. The locks have four sets of machinery, one at each corner, that operate gates and valves. Each year, the crew goes over the equipment at one corner of the lock, so every four years everything gets renewed—in theory, anyway. Electric motors are rebuilt as needed, limit switches and other electric gear are inspected and rebuilt as needed, everything is cleaned and closely examined and anything suspicious is replaced. It's important to do preventative maintenance since replacement parts for some of the century-old equipment have to be custom fabricated. If a gear is broken mid-season and a new one has to be made in the machine shop, boat traffic may be tied up for several weeks. This can be a real problem at the end of the season when the long-distance snowbird boaters from Canada and the upper Great Lakes are heading south to warmer saltwater seas. Approximately every ten years, each lock gets a complete overhaul and close inspection of its critical elements such as doors, pivot and sockets, culverts and concrete repairs. The goal of the canal staff is to do eight overhauls a year, and each rebuild costs around $1 million.

The first step in this process is to install a cofferdam at each end of the lock so it can be completely pumped out. The cofferdam consists of a series of long, narrow steel plates called *needles* placed with a big heavy steel I beam called a *buffer beam* to hold the needles in place against the water pressure. The gaps between the needles are sealed with cinders sprinkled into the water, which sucks them into the leaks. The cinder supply needs periodic renewal throughout a shift to keep the leakage from becoming too great. Once the lock is dewatered, sediment is removed, and work on the valves, gates and concrete structure can begin. During this process, a 24 /7 pump watch must be maintained—sometimes for weeks. The cofferdams always leak a bit, so the pump watch makes sure nothing freezes up that would stop the pump. If a major change in water level were to occur that would overtop the cofferdam, an eight-inch pump on standby would be activated by the pump watch.

The big lock gates are often pulled for maintenance by a crane so that the white oak miter edge where the gates meet when closed can be renewed. Sometimes the crews use big air bags to lift the gates in place to access the pivot and socket so they can be renewed and to access the seal at the bottom of the gates for replacement. The massive valves are sent to the Waterford shop, where they are descaled and painted. Various motors are rebuilt, bearings replaced and babbitted and armatures are baked to remove moisture and then revarnished. Commutators are polished, and brushes are checked or replaced. The electric brake activating rods are also checked and replaced if cracked. All the lock's concrete work, including that of the valve tunnels and culverts, is inspected and sometimes repaired. This work must be completed during the off-season in sometimes bitter cold winter weather. Much of it has to be done in tight spaces, and moving tons of heavy steel gear around has real potential for disaster and damage to human body parts. Sometimes, too, inspections of the dewatered lock can reveal big problems, as with Lock 7 in Oswego. It recently underwent a three-year rebuild at a cost of $28 million. The work included redoing the lock foundation and the rock anchors of the lock walls. Adding to the challenge was that each spring the work had to stop at a point where the lock could be operational for that year's season.

The first time you enter a lock aboard an upbound boat is more than a little intimidating. A certain amount of fatalism is in order as the massive doors astern close with a dull thud that echoes around the slimy, dank, shadowed chamber that you now float at the bottom of. Water spurts in a jet of white through the closed doors ahead, reminding you of the intense forces held at

Right: Lock 17 opening gate with counterweight—note the small blocks on top of the weight for fine trimming. *Author collection.*

Below: Lock 17 lower gate shortly after the Barge Canal opened. *Author collection.*

bay a few yards from your boat. You and your boat are now at the mercy of hundred-year-old machinery designed by a long-dead civil engineer and the training and experience of the lock operator. Then the grind of machinery sounds, and ten-foot boils of water erupt around your vessel as the valves open in the dark water below. The lift begins, and when the step up is twenty or thirty feet, it's surprisingly rapid. Within a few minutes, you have risen out of the cool depths into the sunshine once again.

The ultimate lock experience on the canal has to be Lock 17 at Little Falls, though the Flight of Five at Waterford comes in a close second. Lock 17 has a lifting gate much like those of the guard gates at its lower end. The steel guillotine gate and its massive 150-ton counterweight are mildly terrifying as you pass under the dripping steel slab.

The forty-foot descent of a downbound boat is rapid enough to create the distinct sensation of a gigantic silent flush. Six million gallons of water disappear in minutes. On our first passage east, as our boat dropped down into the dank coolness, dozens of tiny streams of pressurized water came squirting out from pinholes in the wall's steel cladding. This lock's walls are over thirty feet thick at the bottom, and the dripping steel lift gate and counterweight that you must pass under are suspended by century-old mechanisms. It took about four years to build this lock originally, and it replaced four older, smaller locks. Its base was blasted out of bedrock, and it's the only regularly used lock with a guillotine-style gate.

During the opening ceremony in 1916 for the lock, then the biggest lift in the world, the New York governor gave a speech that included this timely

Lockport showing old locks on right next to the two newer locks side by side. *Library of Congress.*

observation as our civilization faces the challenges of climate change and its associated effects: "What is needed today is a vast popular awakening, a realization that government is no automatic device but a tremendous enterprise, and its wise direction is dependent upon the intelligence and interest of every individual."

At Lockport, near the canal's west end, the old flight of five locks beside today's locks is the most complete surviving artifact from the second enlarged Erie Canal. Today's Locks 34 and 35 form an excellent example of early twentieth-century concrete electrically powered lock design. These newer Barge Canal locks pioneered construction methods that were also used on the much bigger Panama Canal. Lockport is unique in having the nineteenth- and twentieth-century structures side by side. Today, the old locks serve as a spillway for the water dumped by 34 and 35.

Waterford: The Climax for Coast-Bound Boaters

The locks that make up the Flight of Five are by far the biggest elevation change on the canal. Here your eastbound boat descends from the Mohawk River into the Hudson Valley, a few feet above sea level. The federal lock at Troy is situated at the beginning of tidal waters. The Flight's five locks have a lift of 169 feet, still one of the biggest canal lifts anywhere in the world. Above the locks, not one but two guard gates protect the approach to the city of Waterford. One gate remains closed except during the operation of the locks. If a lock malfunctioned, the guard gates would keep the stored waters of the Mohawk from roaring down into the Hudson River in a giant waterfall that would sweep much of the city of Waterford away with it. It took ten years to build the locks, dams and guard gates here, and the Flight didn't open until 1915.

Canal locks are perhaps the most conspicuous of the multitude of engineering devices that must work together as a system to keep the waterway functional. It is the experience, dedication and hard work of the various operators who understand the complexity of the system and its vast array of components that keep it all going.

5
A FEW HISTORIC CANAL VESSELS

Few historic artifacts match a ship for being an evocative reminder of the past. Old vessels like the canal tugs *Urger* and *Syracuse*, the motor ship *Day Peckinpaugh* and the *Derrick Boat 8* that once worked on the canal are of special interest to those who enjoy the canal's maritime history. They are part of the world of our grandfathers and great-grandfathers, men of Upstate New York who ran steam engines, laid stone, drove horse teams and worked on our waterways. Many of these men did their jobs with pride. They stoked the boilers of industry with coal, riveted steel for ship hulls and bridges alike and machined the gears and valves that made our canal locks and water-control structures function. *Sine labore nihil*—nothing without work. Work gives purpose. Work with others is utterly necessary to human well-being. And the crew of a ship must work together. Old workboats remind us of how labor on inland waterways helped build our nation and create the living standards we now enjoy.

Derrick Boat 8 and *Dipper Dredge 3*

Derrick Boat 8 now lies ashore at the H. Lee White Museum in Oswego. With its squared-off contours and shack-like wooden deck house, this ungainly riveted-steel barge will never win a beauty contest. Lacking even the dignity of a name, *DB8* nonetheless displays a heritage and history of great import to the world associated with the rise of fossil fuel–powered steam engines.

Derrick Boat 8 now high and dry as part of the H. Lee White Museum collection in Oswego. *Author collection.*

DB8 was launched in 1927 near the end of the age of steam. Already on land, trucks like the 1925 Model T farm truck of my childhood were taking over from horse-drawn wagons, while various internal combustion engines were running pumps, generators and other equipment ashore on farms and in factories and powering various-sized vessels afloat. No boiler, no stoker and no licensed engineer were required to oversee the gas or diesel engine in those times. The streamlined operational needs, instant starting and lower labor costs of internal combustion compared to a steam plant contributed to their rapid spread through industry. However, *Derrick Boat 8* was built by a New York State agency and equipped with a coal-burning 1927 vertical eighty-five-horsepower Ames Ironworks Boiler and a 1927 one-hundred-horsepower, two-cylinder American Hoist & Derrick Co. steam engine to power its lifting crane. Perhaps the state bureaucracies were still skeptical about those newfangled diesels. Quiet, slow-speed, high-torque steam had its advantages after all. About ten years after the derrick boat was launched, its boiler was converted to burn oil rather than coal, but the vessel continued to operate under steam until 1984, when it was laid up. It was eventually transferred to H. Lee White Maritime Museum in Oswego.

The barge's boiler was built in Oswego, so perhaps it was appropriate for the boiler to return to its birthplace, where it is now displayed. When *Derrick Boat 8* was fabricated about one hundred years ago by Ames Ironworks, Oswego was very much a manufacturing town as well as a port. Though

Derrick Boat 8's steam-operated winch with part of the steam engine in foreground. *Author collection.*

the city's heyday for lake shipping was past in the 1920s, it still saw plenty of traffic on the Oswego Canal. Ames Iron Works dated back to 1854, having evolved from an earlier business that built capstans and winches for the Lake Ontario sailing fleet.

Along with the truss work of its prominent derrick, one of the boat's most distinctive features is its set of "spuds." These heavy wooden timbers were lowered into the soft bottom of the canal to anchor the vessel and steady it during work. Sometimes *Derrick Boat 8*'s crane lifted machinery or, after floods, fished out large logs and debris from the canal. Or it worked at dredging, a never-ending task then as now on the canal. Each spud had a small steam engine and winch to raise and lower it. The derrick crane could lift 150 tons, enough capacity for the heaviest lock door in the system, and its rotating mechanism is clearly visible. In its earlier days, the vessel had accommodations that allowed a nine-man crew to live aboard while working in remote areas of the canal. At that time, the company included a captain, fireman, operator, two deckhands, a cook and three watchmen.

Diesel power and hydraulics eventually replaced steam and cables, and several of the dredges now used on the canal employ suction dredging rather than a clamshell or bucket-type excavator. Big pumps push a slurry of mud and water through pipes to a spoil deposit area on shore. This method is effective for moving fine-grained mud or sand.

Dredging is a constant activity on the canal. Its current edition incorporates many stretches of natural waterways, including major rivers like the Mohawk, Seneca, Oswego and Oneida that are fed by a network of smaller tributaries. The creeks and rivers constantly pick up silt, sand and mud from farm fields or construction sites that are deposited as deltas at their entry points into the canal. A single flood event can dump hundreds of tons of sediment into the canal in a few hours. Then the dredges must suck, dip, scrape or otherwise remove the deposits. Without their constant efforts, nature would quickly reclaim much of the canal, transforming it into a long linear wetland. Today the Canal Corporation operates a number of dredges that attempt to maintain an official depth of fourteen feet throughout the system. *DB8* was one of the earliest Barge Canal workers.

Steam-powered *Dipper Dredge 3*, now retired to the Lyons Dry Dock, was another early dredge that dated back to the first days of the Barge Canal. Its steam plant was built in 1909, and it, too, is listed on the National Register of Historic Places. The barge that carries the derrick and dredge machinery dates to 1929 and at 110 feet is considerably larger than *DB8*.

A steam-powered dredge at work during construction of the Barge Canal. *Jim DeNearing.*

DD3 was built in 1910 by a private contractor and dredged Mohawk River mud and debris between Little Falls and Canajoharie. In 1918, the state purchased the machine to help complete the canal. It and several other hydraulic dredges would be pushed or towed to the site by a state tug, where they would go to work cutting the bank profiles or excavating the bottom sludge with a bucket or sometimes pumping the spoils through a pipeline laid to a nearby diked disposal area. Back in the day when commercial traffic relied on the canal, dredging was a 24/7 operation, and the crews often lived on site for the job's duration, which might be for several weeks. They did so in self-contained floating dormitories known as quarter boats that were equipped with bunks and a galley. Their duty was not unlike manning a ship at sea, as they worked their watches without shore leave or a chance to get home.

The quarter vessels have been retired now, thanks to ready road access along much of the modern-day canal. In the 1950s, along with pressure from commerce to keep the canal open, the twenty-four-hour-a-day work

Quarter Boat dry-docked, as it is no longer in use for canal maintenance, with *DD3* in the background. *Author collection.*

schedule also accommodated the old steam plants. Shutting down the boiler fires and then getting up steam again was time-consuming and harder on the equipment, so two twelve-hour shifts of twelve to fifteen men were housed aboard the dormitory boats while on the job using *DD3* or *DB8.*

The survival of *Dipper Dredge 3* and *Derrick Boat 8* is a rarity among vessels of the era. Only a handful of the dozens of vessels and rigs that undertook the construction and maintenance of the Barge Canal and other mega–public works projects on New York's waterways and harbors during the early 1900s remain in existence. Even rarer are examples of early twentieth-century steam-powered equipment. Possibly only one other intact dredge like the steam-powered *DD3* remains in existence today.

Day Peckinpaugh: The First and Last of Its Kind

The *Day Peckinpaugh* was launched 101 years before this was written in 2022. It predated the steam-powered *DB8* by 6 years and retired about 10 years after *DB8* was laid up. It was the first motor ship designed and built specifically for working on the "new" Barge Canal and was initially propelled by a set of Skandia Pacific Oil Engines that were replaced a few years later by more modern diesels. The vessel was a hybrid of sorts designed to be capable of work on open water of the Great Lakes or on coastal salt water while also traveling within the minimal clearances of the canal. This seagoing canalboat spent most of its life in fresh water hauling coal, scrap metal, grain, fertilizer and powdered cement between various ports located on the Great Lakes and the East Coast. However, it did venture on to salt water during World War II, and several sources say it got as far south as Cuba then.

Launched as the *Interwaterways 101* and later named *Richard J. Barnes* before acquiring its current name, the *Peckinpaugh* operated on the canal from 1921 to 1994. It is presently still afloat and lies mothballed near Waterford at the canal's east end. Its fate remains uncertain, but there is yet hope that it might undertake a new career as a traveling educational vessel.

I and others who sailed Lake Ontario in the 1980s often encountered the *Peckinpaugh* as it made regular runs between Oswego and the cement quarry near Picton on the Bay of Quinte. The cement run was the vessel's last field of operation and took it from Canada to Rome, New York, via Lake Ontario and canal. Sometimes it tied up alongside the *Stephen B. Roman*, another larger cement carrier that also ran between the Bay of Quinte and several other Great Lakes ports, including Oswego. Here, the *Roman* sometimes transferred cargo to the canaller. *Peckinpaugh* could carry about 1,500 tons of cement.

It is an odd-looking little ship that appeared squashed thanks to the low profile required by the canal's bridges. It was sized to fit the canal locks at 259 feet with a 36-foot beam and a draft of about 14 feet, the locks being 300 feet long and 43.5 feet wide. While *Peckinpaugh* would not have won any ship beauty contests in its day, it was utilitarian and efficient. History records that on its first trip on the canal, it carried three thousand bushels of oats at a rate 60 percent below railroad cost. And it worked for over seventy years.

The *Peckinpaugh* was listed on the Historic Register in 2005 partly because of its origins as a specialized vessel for canaling. It designer was Captain Alex McDougall, who had previously designed and patented a unique freighter for Great Lakes service called a whaleback. McDougall started sailing the upper lakes as a youth in 1862, and he experienced enough severe weather to acquire a keen interest in and appreciation for seaworthy ship designs. He was willing and able to think out of the box with his ideas and was attracted to the strength of a relatively new method of shipbuilding that used iron rather than oak. His whaleback freighters had well-rounded decks and no

The *Day Peckinpaugh* passed under a guard gate in the Champlain Canal during its tour in 2009. *Duncan Hay.*

bulwark rails. In rough weather, the seas would simply wash right over the vessel rather than collecting on deck and weighing it down. The ships were strong and seaworthy, if odd in appearance, and about forty were launched. However, their small hatches slowed the transfer of cargo, and soon the design was abandoned. One whaleback launched in 1896 operated until 1969, finishing its days as a tanker, and is now a museum ship in Duluth.

Interwaterways 101 was the first of five identical seagoing canalboats, and it went to work carrying bulk cargoes for twenty years until World War II's arrival, when it was drafted to serve the country as a coal carrier. Its mission was to refuel vessels that were traveling the Atlantic in convoy. For several years, wolf packs of German U-boats stalked merchant ships right off the East Coast, and there is a story of a torpedo being launched at the then named *Richard J. Barnes* that passed underneath the shoal draft canaller. Possibly the German captain thought the low-profiled canaller was a heavily loaded coastal freighter and so misjudged the depth of its hull when the torpedo running depth was set for the attack. There is also a tale about one of *Barnes*'s identical sister ships nearly being bombed by a patrol plane that took the low-slung hull for that of a submarine.

After the war was over, the *Barnes* was reconditioned and given a new midships section for cargo to increase its capacity. It went back to work carrying a variety of cargoes on the canal. It carried corn from Toledo, Ohio, via the canal to Seaford, Delaware, a port on a tributary of the Chesapeake Bay. It also transported scrap iron to Brooklyn and fertilizer to Carteret, New Jersey, along with rolls of wire and anything else that was profitable and could be loaded into its holds. Some of its sister motor ships were tankers that carried oil and refined product.

The canal was a busy place in the days before the St. Lawrence Seaway when the motor ship fleet and various tugs and barges were in service. While the *Peckinpaugh* was the first and the last "hybrid" canal/coaster to work on the canal, dozens of others also operated on its waters during the 1920s and '30s, including the Ford fleet. In 1958, the *Barnes* was purchased by Erie Navigation and renamed the *Day Peckinpaugh* and given a set of Detroit Diesels. A few years later, it was converted to a bulk carrier for powdered cement. A new single large cargo hold was installed and equipped with a scraper blade that pushed the cement into a hopper connected to piping that used compressed air to push the powder up into storage silos ashore. It then began transporting dry cement between the recently built quarry at Picton, Ontario, and Rome, New York. The round trip between Rome and the Bay of Quinte took about seventy hours, and the quantity moved on each run was equal to about eighty-two truckloads.

It made dozens of regular runs carrying perhaps 100,000 tons of cement each season and was well known along the Oswego Canal. A woman who waved from her kitchen window, an old man who would blow his bugle for the passing ship and others watched for her appearance each week. One canal-side resident as a boy in the 1960s, who hung around the waterfront, recalled a cook aboard the *Peckinpaugh* who, upon request, tossed the kids a few hot dogs from the galley while the vessel locked through at Fulton. (The kids never got any buns, though.)

In 1989, the raging canal lived up to its name when the *Peckinpaugh*'s crew went to the aid of a sixty-five-footer in trouble on Oneida Lake. The Barge Canal had been rerouted to run through this twenty-mile-long lake, and it wasn't long before Oneida got a reputation for being a nasty body of water. In the early days of the Barge Canal, a number of small old wooden barges were still in service. When westerly gales or nor'easters blew, tugs would sometimes have to wait for days before venturing out with their tows of vulnerable barges. It was bad enough that the state actually acquired a tug to serve as a rescue boat in the 1930s. It was based at Sylvan Beach on the lake's east end and rescued dozens of barges from being grounded and broken up on the lake's eastern shore. One news story cited in the book *A Long Haul* recalled a fifty-mile-per-hour gale and "mountainous waves" that sent a fleet of barges loaded with $240,000 of wheat onto the beach. The rescue tug *National* managed to salvage the barges and assisted their crews in that storm.

The *Peckinpaugh*, designed for Great Lakes waves, had little difficulty with Oneida Lake's steep, short chop, but during an October blow, a Florida-bound vessel, the *Seabreeze*, battered by the shear, choppy waves sent out a distress call. The *Peckinpaugh* crew was able to lower their lifeboat and got three of the *Seabreeze*'s four-man crew off. The last was rescued by a state police helicopter after the *Peckinpaugh* radioed the lock tender at the lake's east end.

In 1994, Erie Navigation lost its contract for cement, and the *Peckinpaugh* made its last run. It was then laid up in Erie, Pennsylvania, for more than a decade. In 2005, interest in saving the vessel as a historic artifact and traveling classroom on the canal sparked its rescue from an imminent fate of being scrapped. The rescuers envisioned a larger version of the tug *Urger*, which had traveled as an ambassador boat and educational resource on the canal for over twenty years. With cash from a devoted Canal Society member and others, the vessel was acquired for the New York State Museum with the goal of turning it into a floating and traveling educational exhibit. It was towed from Erie to Lockport to be refurbished by volunteers and in 2009 set forth on a "proof of concept" voyage north to Plattsburg via the Champlain Canal and south to New York City.

John Callaghan, a New York Canal Corp employee and historic consultant, wrote during the 2009 cruise: "She carried history, imagination and hope....A century ago the 'Day Peckinpaugh' represented a quantum leap forward in maritime innovation. Today, she endures as an important link to our past, and a reminder of what is possible when the right people and resources come together for a common goal."

Since then, the *Peckinpaugh* has again been laid up and now lies in Waterford, once more facing an uncertain fate. Historian Duncan Hay, author of the Barge Canal's National Historic Landmark status as a "historic corridor," wrote in 2021 for the Canal Society of New York's publication *Bottoming Out*:

> *Let's hope this isn't the end of the "Day Peckinpaugh" saga. She's a tough old girl who has gone through hard service, multiple engine replacements and a couple of major rebuilds over the past century, She's been "re imagined" several times and escaped near-death experiences*

because she was solidly built, practical and adaptable. Beyond that, for reasons I don't fully understand, that homely old boat has captured the hearts and imaginations of countless people as she chugged across Upstate New York and the Great Lakes.

TUG *URGER*: A CENTURY OF WORK

The tug *Urger* is special. It was placed on the National Register of Historic Items in 2001. Until 2018, it may have been the oldest fully operational work boat afloat in North America, and many of the men and women who operated her had great affection for the old girl. It is a survivor from an earlier age when coal, steam, polished brass and riveted steel reigned supreme. *Urger*'s classic tugboat lines and rugged simplicity appeal even to landlubbers. "You can't see her and not love her," one of her crew told me a few years ago.

It was built in 1901 in a Michigan shipyard and was purchased by New York State for canal duty around 1920. For more than eighty years, *Urger* traveled the state's canal system pushing and towing barges and dredges to and fro. It was retired from this work in the 1980s. However, after a short hibernation, it was revived and put back to work as a "teaching tug" and was visited by thousands of schoolchildren while traveling the canal system over the next twenty-five seasons.

The *Urger*'s history parallels that of smokestack America and its culture, society, natural resource exploitation and changing transportation and energy industry. It was built when the Great Lakes region's industrial economy was ramping up to become the mightiest producer of machinery, tools and twentieth-century transport infrastructure in the world. It and the New York Canal System belong to a muscular time of steel, steam and highly skilled craftsmanship. Steam power reigned supreme, and coal was king when *Urger* was launched. At this time, water, iron ore and coal fueled the vast Great Lakes manufacturing complex of foundries and factories, mills and makers of various products that stretched from Wisconsin south to Chicago and Gary and east to Buffalo and Hamilton, Ontario. When the tug's steam plant was installed, electricity was still in its infancy as a source of industrial power. The first big alternating current hydro station in the country, located at Niagara Falls, had been operating for only a few years. Steam still ran the world, though not for much longer.

The *Urger* was designed originally as a fish tug for work in the gill net fishery of Lake Michigan. Its crew set and hauled gill nets and carried the fresh catch ashore to processing and freezer plants. The fishery in which it worked was at that time one of the greatest and most profitable freshwater fisheries anywhere on earth. In 1901, millions of pounds of whitefish, lake trout, lake herring and sturgeon were landed each year on the Great Lakes, and thousands of people worked at catching, processing and distributing lake fish throughout the heartland of America.

The tug was given a sturdy steel hull to withstand the heavy ice of a winter fishery. Its coal-fired steam plant was installed by the workers of the Johnston Brothers Shipyard and Boiler Works, a company that to this day still builds Scotch marine fire tube–type boilers for ships. The *H.J. Dornbos*, as it was christened, was seventy-three feet overall and about forty-five tons displacement. A contemporary news account called it the "finest boat" in the Lake Michigan fleet. But already, even as the tug first hit the water and was christened with its owner's name, the fishery was hinting at its finite future.

The *H.J. Dornbos* as a steam-powered fish tug. *Author collection.*

By 1900, the Great Lakes fisheries were quickly transforming into the industrialized trade we see today. Small sail-powered wooden fishing skiffs that had served gill netters since the beginning of the trade were giving way to larger iron- and steel-hulled fish tugs. Steam-powered tugs could go farther offshore, carry more fish and larger crews and set and retrieve far greater lengths of gill net with their powered net lifter winches. This, along with improved transport and freezer plant technology, resulted in a rapid expansion of fishing effort, landings and profits. From 1880 to 1890, the number of steam-powered fish tugs tripled on the Great Lakes, and income from the fishery shifted to and concentrated in the hands of wholesalers rather than the independent fishermen. Soon the familiar pattern of overfishing and subsequent collapse of overexploited stocks followed. Even before the *H.J. Dornbos*'s launch in 1901, some people had written about the decrease in size of the individual whitefish and trout being caught, a classic sign of an unsustainable fishery.

As fish stocks declined, smaller, more economical vessels with internal combustion engines rapidly made inroads in the fishing business. The *H.J. Dornbos* ceased to be profitable and was sold in 1910. As the more appropriately named *Urger*, it then towed and pushed barges around Lake Michigan for about ten years. Around 1920, it was purchased by the New York Department of Public Works for use on the recently expanded Barge Canal.

Urger was kept busy urging barges of riprap, dredge spoil and loads of various canal gear around the state to improve and maintain the new waterway. By the 1960s, however, the St. Lawrence Seaway had opened, and by the 1980s, traffic on the canal had become a fraction of the movement of thirty years before. *Urger* was no longer needed. It was laid up and faced a possible scrapping. However, a successful businessman and World War II navy commander, Schuyler M. Meyer, had other ideas for the old canal worker. Meyer had a long-standing interest in both maritime history and education for children and used his own funds to operate *Urger* as a teaching tug for several seasons. He toured the canal system with it as volunteer captain of a floating classroom to demonstrate the feasibility of a tug as teacher, before the state took over the program.

Urger was named to the New York State and National Register of Historic Places because of its age, long service and several design features. Its steam plant was replaced after World War II with a war surplus diesel engine, a slow-speed six-cylinder Atlas-Imperial. The engine weighs nearly nineteen tons and is a model used in a number of wartime minesweepers. Atlas-Imperial engines were considered one of the most serviceable diesels ever

Urger tied up in Oswego around 2012. *Author collection.*

built in the United States. The Oakland, California company made diesels for fish boats, tugs, coasters, yachts and other ships. Their engines came in a variety of sizes and ratings, from a two-cylinder model that generated thirty horsepower to an eighty-cylinder model that generated six hundred horsepower. Known for their reliability and durability, Atlas diesels were installed in workboats around the world. Some of the old engines outlived the company, which closed in the 1950s. *Urger*'s diesel was one of a few dozen such motors still in service. The 6 H M 1558 model shoved the old tug along at an economical, fuel-stingy forty revolutions per minute (rpm) year after year. Its top speed is a little over three hundred rpm. (Today's noisy little diesels that pound along to power many sailing yachts and slow-speed power craft typically run around two to three thousand rpm.)

The engine is a direct reversing type, meaning it has no transmission but rather must be stopped and reversed to back up. There is no neutral, so to stop, the engineer simply shuts it down. It is then started up in reverse with compressed air supplied by a second smaller engine running a compressor. *Urger* is a "bell boat," meaning that the captain signals the engineer by a series of signals that ring bells on the forward bulkhead of the engine room. This is an antiquated system dating back to the earliest days of steam navigation on the lakes, and few boats were still operating with such signals in the twenty-first century—especially boats that lacked a neutral gear. Successful docking or other tight quarters maneuvers are, as one crew member put it, a delicate dance requiring a high level of trust between the wheelhouse and the engineer.

After more than a century of nearly continuous service, the *Urger* needs extensive repairs to continue working. This will be expensive, and those who respect and admire the old vessel fear that the Power Authority will not revive the educational programs that kept *Urger* chugging along the canal for many years. If the old boat is cut up for scrap, a piece of unique American history will be lost forever. As of this writing, it was on the hard at the canal maintenance facility in Phoenix. A marine survey suggests the old boat is in need of a new bottom, and such a job will not be cheap, so like the *Peckinpaugh*, *Urger*'s future is uncertain. Yet this unique and historic vessel has much to teach us about the past.

Historic artifacts can be evocative, and for many people, ships and boats are especially so. Stand for a moment on the catwalk of *Urger's* engine room or in its pilothouse and all sorts of thoughts come. You think of movement and work. A tug like *Urger* represents an unbroken lineage of maritime skill and knowledge. *Urger*'s life spanned the rise of industrialization, the high point of post–World War II American influence and the subsequent deindustrialization and shift to globalized trade. *Urger*'s gleaming brass and copper fittings and starkly simple pilothouse evoke nostalgia. *Sine labore nihil*—nothing without work. Work gives purpose. Work with others is utterly necessary to human well-being. And the crew of a tug must work together. Especially if their engine room runs on a bell signal system. *Urger* and the other historic vessels of the canal remind us of how waterways helped make our nation. There are maritime museums of yachting that preserve the polished artifacts of the gilded golden era. Workers like *Urger*, *Peckinpaugh* and *DB8* created that wealth.

Keeping an old canal tug or motor ship operational would be a suitable memorial to the countless workers of the state's factories, shipyards and waterways. These vessels are memorials to men who labored on the canal and to others, male and female alike, unsung and simple who believed in the future of New York State and their country. Hopefully, *Urger* will receive the necessary repairs and upgrades that would allow it to continue its educational programs for years to come.

These historic vessels represent an unbroken lineage of skill, work and knowledge that reaches back to a day when America was young, largely rural, but reaching toward the peak of its powers. Their fates, like our own, are uncertain in an age of climate change, as we face an urgent need for a reimagined transport system in a post–fossil fuel age. Might that system include a revitalized canal system with vessels powered by batteries charged by solar power?

6

STATIONARY ENGINEERING

Guard Gates, Spillways, Bridges, Hydro Power Plants and Night Navigation

Along with locks and dams, a critical feature for water control on the canal is the appropriately named guard gate. As the name implies, its function is to protect the canal and adjoining lands from flooding in the event of a catastrophic equipment failure. There are about a dozen guard gates stationed at various critical points along the canal that can be closed to isolate a section of the waterway. They generally are situated where a line-cut section of the canal meets a natural waterway that could allow a dangerously large amount of water to enter suddenly during a flood event.

The steel gate is usually left open so traffic can pass under it, except above Waterford. Here, two gates protect the city from a lock failure and sudden flow of the Mohawk River down the incline. They are opened periodically for boat traffic. In the down position, the gate fits into a sill on the canal bottom similar to that used at a lock door. There is also one guard lock where the canal crosses the Genesee River. It has a regular lock chamber but is equipped with guard gates rather than conventional miter lock doors. The lock maintains the navigational pool when the Genesee River level is high or low. Rarely used since 1952, when the construction of the Mount Morris dam upstream stabilized the river's flow, the two gates prevent the canal from draining into a drought-stricken river or from being flooded by it.

The line-cut sections of the canal are drained each fall and filled again in the spring. This is done for maintenance on the dams, locks and other

Guard gates at Rome—in an emergency, the machinery on the steel bridge over the gate allows the gate to be lowered, controlled by cables and counterweights on each side. *Author collection.*

Spillway in Oswego with open lock in background. *Author collection.*

SPS *Salem*, a self-propelled scow formerly used to set buoys—note the swivel drive for maneuvering. *Jim DeNearing.*

Dry dock miter gate to allow boats in and out. *Author collection.*

Another dry dock view showing the blocking with tug and scow. *Author collection.*

structures as well as to protect the waterway and its equipment from ice damage and seasonal floods. Each fall, the guard gate at the canal's west end is lowered to block Lake Erie water as the line-cut portion of the canal is drained. In the spring, it's raised to fill the canal.

The spillway is a prosaic and inconspicuous feature of the canal that controls high water levels. It's simply a low dam with a concrete slope that, where the canal crosses a creek or runs along a river, allows water to spill into the natural waterway channel below when canal pool levels are high. Spillways are rarely visible to the land-based traveler. When you're on a boat and first see a spillway, it appears as a gap in the canal bank, and the sudden space where normally a solid bank exists is vaguely alarming, especially if water is passing over the lip. Down below, the creek bed receiving the flow is sometimes littered with huge logs and tree stumps, suggesting the occasional power of unwanted water on the canal. Once while walking along the Oswego River, I looked over at Lock 7 and saw a family of geese taking a shortcut from the canal to the river. The birds slid one by one down the concrete apron of the lock spillway that was dumping water from the recent operation of the lock to reach the river.

The weir is another structure unique to the canal and also often overlooked. Its function is to add water to the canal. Like the spillway, it's simply a low dam that allows water to flow freely over the top into the canal but makes no attempt to store the water in a reservoir. Spillways also allow easy measurement of the volume of water entering the canal, something the system operators need to monitor constantly.

One of the most intriguing types of water control structures on the canal (or anywhere else) is the dry dock, an area where boats are maintained and stored "on the hard." The canal maintenance facilities in Lyons and Waterford each have a dry dock. The docks are large basins connected to the canal by a narrow passageway equipped with two miter gates similar to those of the locks. A series of small passageways with valves controls the flow of water into the dry dock so that boats can be floated in and positioned over the support structures. Then the gates are closed, and the dry dock is drained dry. The boats are lowered to rest on the cement blocking and secured. In the spring, the various vessels are refloated. The Lyons dry dock has a permanent resident, *Dipper Dredge 3*, that has been listed on the National Register of Historic Places.

Getting a Lift on the Canal

Along with lock dimensions, bridges defined the new Barge Canal's limits of capacity. During the design phase, there was considerable discussion about vertical clearance heights as the possibility of building a canal to allow passage of ships from the Great Lakes was debated. Ultimately, cost considerations prompted the eventual design of a minimum clearance of about sixteen and a half feet for barge traffic. In the 1930s, federally financed improvements on the canal east of its junction with the Oswego River raised clearances to around twenty and a half feet. (Clearances of fixed bridges can vary by several feet as the canal water levels change throughout the system.)

On the western section of the canal that follows the route of the original Erie Canal line cut, the various canal-side towns and villages have lift bridges. There are sixteen in all, and in their lowered position they line up with the road grade, typically just a few feet above the water. At each end, machinery pits accommodate the lifting mechanisms of cables, sheaves and cast concrete counterweights. The weights sink into the pit when the bridge truss is raised. One end of the bridge always features a bridge tender's

Fixed bridge on Oswego River. *Author collection.*

control tower with a set of stairs leading up to it. Electric motors raise and lower the spans.

During the canal's busy commercial era, each bridge had its own staff, and the operators were stationed throughout twenty-four hours. Today, like the locks, they generally operate only during the day. Most of the bridges these days operate on demand, and anything much bigger than a canoe or a rowboat needs a lift. The boat skipper calls or radios ahead to announce his or her pending arrival because many of the operators "rove," driving between bridges to accommodate boater traffic. The bridge tenders also keep in touch with each other and will often notify the next operator down the line about the approach of a through-traveling boat. During a trip from Oswego to Buffalo that we made in 2019 on our small sailboat with the mast on deck, we never waited more than five or ten minutes for a lift.

The bridges today are generally pretty reliable, but now and then they do get stuck. It's usually not for long though before repair crews get them operating again. It takes constant maintenance and plenty of funding to keep them reliable. The busy Fairport lift bridge was closed for nearly two

Top: Lift bridge in raised position with second one lowered. *Wikimedia Commons.*

This page, bottom: This bridge at Lockport built in 1914 was claimed at the time to be the widest bridge in the world. It is 475 feet wide and spans 116 feet of the canal. *Library of Congress.*

Railroad Bascule (aka jackknife) Bridge on Tonawanda Creek built in 1918—note the counterweight. The bridge could lift up from one side to allow ships from Lake Erie to pass. *Chris Gateley.*

years starting in 2019 and left in the up position for major repairs to the deck and lifting mechanism that cost over $16 million. Some of the cost and delay was due to getting replacements for parts of a 107-year-old 340-ton bridge. Fairport's bridge is unique in that it was built with a thirty-two-degree skew from perpendicular across the canal, and its deck has a four-degree incline.

HYDROS

Hydroelectric facilities associated with the canal generate a little over 2 percent of the state's total hydropower supply, contributing approximately $24 million of electricity annually to the grid. Hydroelectric production on the canal was part of the new waterway's design from the very beginning, as it was essential for operating the lock gates and machinery. The original design called for most of the locks to have their own power plants using the water's flow through the locks to generate electricity. Additional hydroelectric stations were also installed along the Oswego, Genesee, Mohawk and Seneca Rivers. Today, a total of twenty-seven hydros from three- to ten-megawatt (MW) capacity operate along the waterway. They include the oldest continuously operating hydro in the country. Some are operated by the state; others are privately owned.

Historically, many nineteenth-century mill towns on rivers that are now used by the canal adopted hydropower for their factories at about the time the new canal was being planned. The hydropower plant at Lock C-2 at Mechanicville along the Champlain Canal was built in 1897.

It supplied power to General Electric facilities and to the American Locomotive factory in Schenectady. The original generators are still producing electricity and profits today, more than a century after startup. It is the oldest known continuously operating three-phase power plant in the world. It was extensively renovated in 2003, and according to Wikipedia's article on it, in July 2021, it began providing electricity for bitcoin mining.

The privately owned 38-MW hydroelectric plant on the east end of the Mohawk River at Cohoes uses the flow of one of the largest waterfalls by volume in the state. Its GE turbines built in Schenectady over a century ago keep generating enough power to light up to sixteen thousand homes, proving the longevity of well-maintained, low-speed, low-temperature turbines and

Generator in Oswego at Brookfield Plant, one of three in the building. *Author collection.*

Mechanicsville generators around 1897. *Wikipedia.*

generators. At the canal's west end at Lockport, a small station operated by the Canal Corp. generates a maximum of 3.5 MW from the flow of water down the Niagara escarpment into the canal via a tunnel around the locks.

Some of the plants along the canal are owned by the Power Authority, the corporation that also operates the canal. Others are owned by independent power producers like Brookfield Renewable Energy, a multimillion-dollar international corporation that runs hydros all over the United States.

The facility at the Crescent Dam next to the Waterford Flight generates about ten MW at peak flow. It, too, includes century-old equipment refurbished to make electricity for the New York Power Authority. Several hydros operate beside dams on the Oswego River. The dam near Lock 7 has a small hydro that is typical of a number of stations along the waterway. Here three generators hum away driven by turbines that are fed by the Oswego River. The river supplies up to 2,200 horsepower, but in the summer, its flow often diminishes to the point where stillness prevails in the powerhouse. The Federal Energy Regulatory Commission (FERC) regulates how much water the small hydros can divert from the river. A minimum flow must be left for the welfare of the fish and other aquatic life downstream from the dam. Sometimes during dry summers, none of the river flow can be used for power production.

Cohoes Falls beside the Waterford Flight. The seventy- to ninety-foot-tall falls span a one-thousand-foot width. *Wikipedia.*

Many of these little stations have century-old equipment that gets a rebuild every thirty to fifty years. In a day of disposable everything, it seems little short of miraculous that such low-tech equipment can remain profitable. But it is. In 2022, Oswego began renovations on its city-owned High Dam and associated hydro plant to allow increased power production. The plan was to spend $4 to $6 million on getting the station back to full capacity. About $1 million was allocated to repair the dam spillway, while another $2.4 million was being spent to modernize the power production process. At peak river flows, the plant will then be able to produce enough power to meet all the city's residential demand, avoiding 24 million pounds of potential CO_2 release into the atmosphere while doing so. At 100 percent capacity with peak flows of water, the plant will be able to produce $1 million of electricity every month.

Navigation and Nightlife on the Canal

Even in the earliest days of commerce, traffic on the canal moved at night. When the canal was a simple line cut that was forty feet wide, it wasn't too difficult—even with only a feeble kerosene lantern hung on the bow—to

Lighthouse at Verona Beach built in 1915, electrified in 1927. *Author collection.*

keep track of the boat's relationship to the bank. But when the new Barge Canal opened, there was no towpath. Rather, the channel followed the bends and twists of natural rivers and the expanses of lakes for much of its length. Sometimes tributary streams, bays or islands complicated the navigational picture, so buoys were needed to mark the channel or hazards near it. Oneida Lake's twenty-mile length and open waters featured a number of shoals and rocks, so here channel markers were critical, and the canal entrance on each end of the lake had to be marked by a lighthouse.

The state adopted the navigational standards used on the seacoast (originally black, red and white buoy colors) and initially placed more than 1,500 aids to navigation throughout the canal system. Many were lighted for night work. In 1919, another 100 lighted buoys were added, and as night travel increased, still more were installed. Traveling 24/7 reduced transit time considerably—a round trip from New York City to Lake Erie and back could now be done in two weeks, while the run from Troy's tidal waters to Buffalo could be done in less than five days by a tug and barge.

In 1918, lighted markers were provided with kerosene lamps. Their large reservoirs of fuel allowed a burn time of a week or more, depending on the weather. Then the lamps had to be refilled, wicks trimmed and other maintenance done. This task of looking after more than two thousand

A small portion of the channel markers that are placed each season on the Oswego River. *Author collection.*

Buoy boat at Camillus Canal Museum. *Author collection.*

channel markers, many of which were lighted, fell to a fleet of buoy boats and their operators. The buoy boat was typically powered by a low-rpm heavy-duty gasoline engine and was twenty-eight feet in length with a beam of seven feet. It carried two thirty-gallon galvanized tanks on the afterdeck, one for kerosene to fuel the lamps, the other for the engine. The fleet was designated by BB plus a number, and as their operators patrolled the canal, they would also take note of other navigation hazards, including snags, bank erosion or other issues such as a marker off station or otherwise needing attention.

The Camillus Canal Museum has a boat on display, number 159, launched in 1930. It, like a number of its sisters, was built by the state. This boat was powered by a three-cylinder Lathrop engine that typically used fifteen gallons of fuel for a day's work. The solo operator usually changed thirty lanterns a day. He carried ten lanterns on the foredeck, lit them and exchanged them for the ten lanterns that were mounted on the buoys. He then refilled the ten exchanged lanterns, trimmed the wicks, relit and exchanged them for the next ten lanterns.

7

CANAL HYDROLOGY

Keeping the Water Where It Needs to Be

The canal system today is a complex network of rivers, lakes, reservoirs, feeders and other waterways. It takes more than two thousand man-made structures to keep the water where we want it. If it's too low, boats run aground. If it's too high, they can't get under the bridges. Then there are shore-bound canal users to consider. Hydroelectric stations want as much water as they can get to turn their turbines and generators. Farmers need access to irrigation supplies. Environmental considerations and needs such as those of federal wildlife refuge wetlands like the big Montezuma marsh complicate canal water management. And the threat of flooding is a constant headache. In a few hours, a bad flood can do tens of millions of dollars in damage. And sometimes floods are lethal to human life.

Water management, as one Canal Corporation document puts it, is an "arduous" process. Today, as in the past, competing interests sometimes clash. When the second enlarged version of the Erie was built, the brisk traffic of the 1850s constantly locking up or down taxed its supply of water. Then there came a dry summer, and the village of Skaneateles delivered an ultimatum. That year, Samuel Hopkins Adams wrote in his book *The Erie Canal*, the villagers observed their water flowing at a rapid pace out of the clear depths of Skaneateles Lake to feed the canal. The townspeople gathered up gunpowder from their homes along with old bolts, nails, screws and other iron scrap. They then went off to the town square to transport their war memorial, a War of 1812 cannon, to the Lake Outlet. Here they

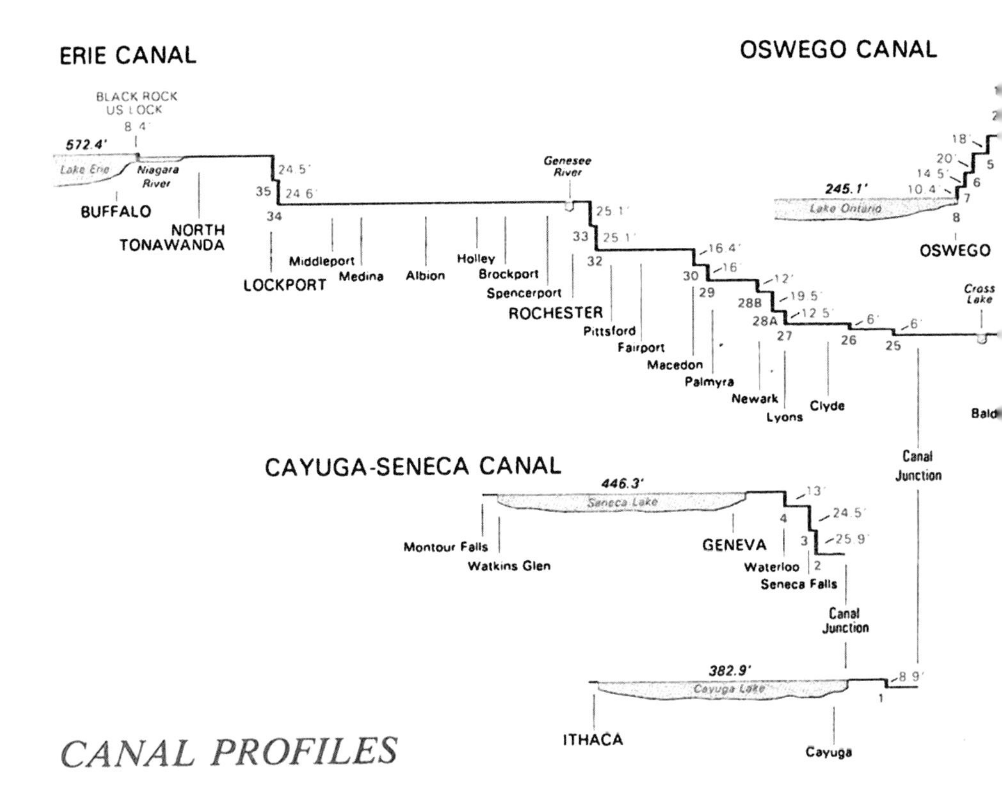

CANAL PROFILES

shut off the water and hauled the cannon to a commanding view of the sluice gate. Soon the Canal Commission engineers arrived to find out what happened to their water supply. The spokesperson pointed out the cannon and shrapnel and the several shotguns that citizens had brought along to the meeting. "You will find blood and thunder in Skaneateles, gentlemen, if one of you touches those gates." Apparently, Adams wrote, some sort of agreement was then made about encroachments on the lake supply.

Water management challenges on today's waterway are also continuous and sometimes a little contentious during both the navigation season and when winter maintenance to locks and other structures is ongoing. Some interests along the waterway like the water high; others with low-lying waterfront property feel otherwise. Keeping a more or less constant level takes a considerable staff, no small amount of computing power and a vast network of data collection equipment and analysis of same. Adding to the challenges of keeping the waterway navigable is a greatly reduced tolerance for flooding in the twenty-first century even as climate change is making severe floods increasingly likely. Ongoing pressures for residential

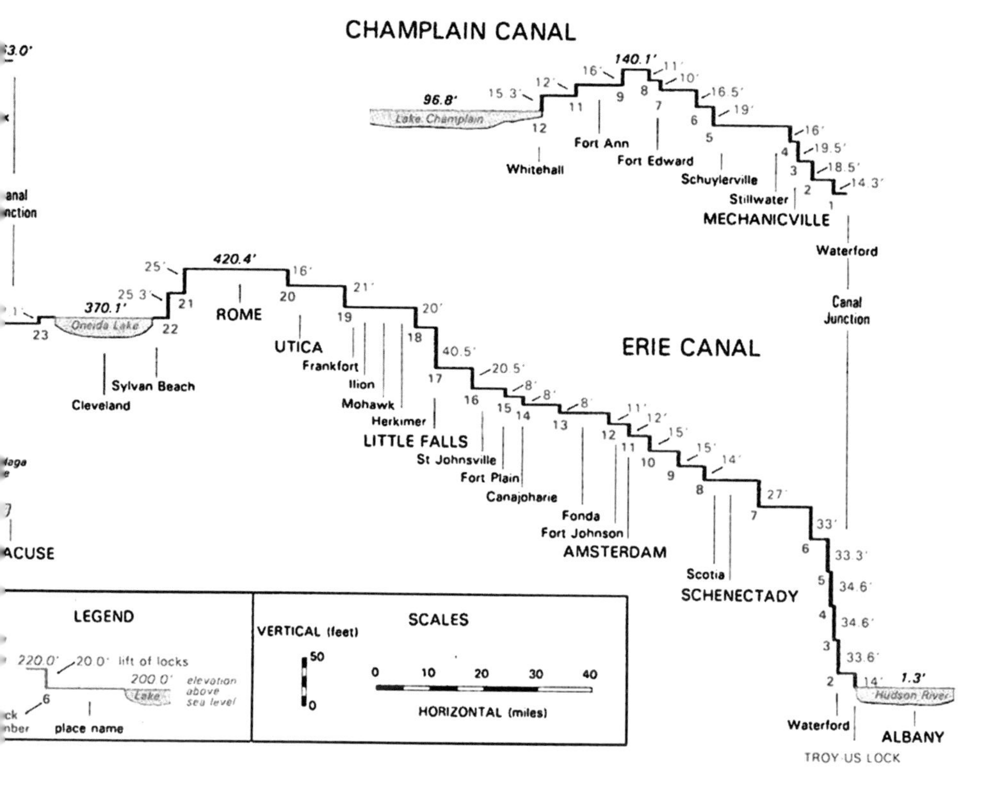

Canal elevations diagram. *NYS Archives.*

real estate development on and near the waterway, the filling of marshes and swamps and deforestation have all reduced the resilience and ability of the surrounding floodplains to soak up and divert water. Still another complication for canal water management is that the levels of the Finger Lakes and Lake Erie are controlled by forces separate from those of the reservoirs that feed into the canal.

The canal's waters are replenished at several places, two of them being at the western end where Lake Erie flows in via the Niagara River and at the canal's "summit" near Rome. Around Lyons in Wayne County, the canal abandons the channel of the Clyde River and becomes a line cut. The Seneca and Cayuga Canal is supplied with water from the two largest Finger Lakes and feeds into the Seneca River, and the Oswego Canal draws on the Oswego River's watershed. The long eastern section to and including the Mohawk requires water from two large man-made reservoirs, Hinkley and Delta Lakes in the southern Adirondacks, plus a feeder that uses part

West Canada Creek feeds into the Hinkley Reservoir just upstream from this location at Prospect Gorge. *Wikipedia.*

of the old Black River Canal. North Lake, another reservoir, in the land of loons and Adirondack spruce, hemlock and pine forests, today serves as the headwaters of the Black River. It was created to serve the Black River Canal and still supplies the present-day waterway.

As early as 1873, extensive logging in the mountains was a worry to those in charge of the canal's water supply. In 1892, the Adirondack Park's boundaries were declared by New York State in large part to preserve forest lands and protect the area's ability to store water to replenish the canal. Today the "blue line" of the park encompasses about six million acres of private and public lands. Not all of it is pristine wilderness, but a substantial second-growth forest does cover the steep slopes in many areas and helps filter and absorb rainfall to maintain rivers and creeks even as the trails, campsites and mountain lakes of the "Dacks" support a vibrant recreational economy.

To the south, the Jamesville reservoir near DeWitt supplies water as a navigable feeder to the present canal, as do several other small reservoirs in Onondaga and Madison Counties. Some of the water flows through a portion of the nineteenth-century second enlargement of the Erie Canal that once ran south of Oneida Lake east of Syracuse.

The canal and its connecting waterways are part of an incredibly intricate and interrelated system of surface, atmospheric and subsurface water—all of it restless and on the go. Few people are aware that what happens in a river can affect unseen water supplies beneath their feet. Even fewer would suppose that the canal's level could have an effect on their drinking water supplies. But that was made dramatically clear in the summer of 2020 at the canal's east end.

Underneath rivers and streams, there exists a watery layer of sediment and stone called the hyporheic zone. It extends considerably beyond the banks of the surface water and often merges with groundwater, especially when stream flows are high. This underground water also flows, though much more slowly than does surface water. As it moves along, it filters and purifies water, helping keep the stream healthy. Some have termed the hyporheic zone the "liver of the river." Microbes and small invertebrates live within the water-filled spaces of the zone's sand and gravel and contribute to the stream's productivity.

In 2020, because of the COVID pandemic, the opening of the canal was delayed by several months. It was a dry year, and the Mohawk River's water supply dropped dramatically without the canal's feeders, dams and locks keeping levels constant. Large stretches of gravel and sand marked the usual bottom of the river during summertime canal operation. Various small businesses including waterfront restaurants, kayak and boat rental operations, marinas along with anglers had to cope with the lowest water seen in many a year.

As the river dropped, it also affected the Great Flats Aquifer beneath its hyporheic zone. The aquifer is the sole source of drinking water that fills the municipal wells of several cities in the Albany area. As the aquifer level dropped around the wells from withdrawals by the cities, various toxic chemicals, including toluene and trichloroethane (TCE)—along with some legacy radioactive contaminants from past military research indiscretions north of the river—began moving toward the municipal well fields. Disaster was averted by the canal's belated opening in July, which allowed aquifer recharge from the Mohawk to bring groundwater levels up to their normal range.

Floods: Four Hours of Warning Can Save Millions of Dollars

In recent years, a half-dozen floods on both the Mohawk and Oswego have shut the canal down and damaged equipment, costing millions of dollars to repair. One of the worst was in 2011, when Hurricane Irene moved up the coast in late August. By the time it reached the capital region, it had been downgraded to tropical storm status. But the storm caused an estimated $13.5 billion of damage throughout its U.S. run, making this one of the costliest storms ever. Two weeks later, Irene was followed by Tropical Storm Lee. In eastern New York, entire villages were rendered uninhabitable after 13 inches of rain fell, and 55 stream gauges recorded new records. Schoharie Creek at Burtonville was 675 cubic feet per second on August 27 and 128,000 cubic feet per second the morning of August 29. This, a local news story noted, was more than the mandated minimum flow over Niagara Falls during tourist season. The flood elevation peaked at 16.8 feet at Lock E10 more than 3 feet above the 500-year flood elevation of 13.5 feet over the normal baseline for the Mohawk. Among the casualties of the flood was a covered bridge dating to 1855 that was one of the longest single-span covered bridges in existence.

One reason the flood caused so much havoc was uncertainty about the predicted track of the storm. As Irene approached, canal workers tried to draw down the level of the Mohawk by raising the top gates of the movable dams. But by the time it became clear that Irene would track inland enough to have a massive effect on the river system, there was too much water pressure, and it was too late to move the lower dam sections out of the river's flow. Contributing to the issue of too much water, a precautionary effort to reduce pressure on the Gilboa Dam, located on a tributary of the Mohawk, was underway. Workers there were dumping water from the Schoharie Reservoir, part of the New York City water system, to stave off a possible dam failure. Adding to their own sense of urgency was the possible effect on the dam of a recent mild earthquake of 2.9 magnitude that had occurred the day before the rains began.

Flood damage to the canal locks and movable dams was heavy. At Lock E9, the river detoured around the dam, carving out a small canyon to its north, and the powerhouse at Lock E10 was so undercut it fell into the river. Other locks experienced similar undercuts, scouring and damage. At Lock E8, a barge crashed into the gates. Five of the eight movable dams were damaged by the flood. At Lock E10, the stone Guy Park Manor house on the canal's bank nearly washed away.

Blenheim Covered bridge before it was destroyed in 2011. It has since been rebuilt. *Wikipedia.*

Schoharie Creek Aqueduct was demolished when river canalization was complete. *Author collection.*

Georgian-style manor house built in 1774 for Guy Johnson, stabilized but still showing damage from tropical storm Irene. Photo taken in 2020. *Author collection.*

The house, listed on the National Register of Historic Places, was built in 1774 for the son-in-law of Sir William Johnson, the famous British military officer who held the alliance with the Mohawks together in the run-up to the Revolutionary War. Guy succeeded his uncle, who died in 1774, but the younger Johnson, a Loyalist, fled for Canada after the Revolution began. His house was confiscated and served as a tavern and stagecoach stop on the Mohawk turnpike for years. The cut-stone structure was badly damaged in the flood, with a whole corner being washed away. Stabilized and boarded up, its future ten years after the flood remains uncertain. It's been proposed to elevate the entire building to prevent future flood damage, which sounds like a pretty pricey endeavor to this writer. Perhaps if Johnson had been a supporter of the rebel cause in 1776 rather than a Tory, his house might have received more funding in the aftermath of the flooding.

By heroic effort, the canal workers managed to make emergency repairs in a few weeks to the locks and dams so as to get dozens of stranded long-distance boaters plus a couple of small cruise ships on their way to the coast. That year, the canal stayed open until December to allow travelers to reach salt water. The 2011 high water event cost the Canal Corp. more than $40 million in damages to the movable dams and other equipment. Full repairs, plus upgrades including reinforced and rebuilt canal dams and improved monitoring to reduce the impact of future disasters, took years and ultimately cost more than $100 million.

Today, snowpack data throughout the canal's watershed, some of it collected by canal workers, is fed into models and helps predict spring floods with far more accuracy than in the past. The National Weather Service's Spring Flood Potential Outlook and modern stream gauges provide real-time data on both river levels and flows, so canal hydrology staff have access to far more timely and detailed information than previously. They in turn alert the lock operators about actions needed to prevent damage. It's a big improvement over simply observing the water levels as shown by a painted gauge at the individual lock walls, as had been the practice before 2011. But it won't be any easier in the future to keep things under control as climate change–related weather extremes keep ramping up. And even though forecasting of large-scale precipitation and subsequent floods has improved, smaller localized downpours can still affect the canal and pop up with little warning. Constant updates and assessments of vulnerability are key to keeping the waterway operational.

However, real estate development and zoning on and near the waterway are generally under the influence of highly localized jurisdictions, as are zoning variances. People like to live near the water, and the zoning variance process is subject to plenty of political influence in many towns and villages. Insurance companies have some influence when it comes to deciding whether a new residence by the water gets built, but comprehensive statewide planning and regulation of floodplain building is, as yet, far from rigorous and effective. It's a good bet that canal hydrology and water level management won't be getting any simpler or easier in the years to come.

8

INVASIVE ALIENS GO CANALING

Nonnative species deliberately or accidentally imported have repeatedly devastated native ecosystems since the arrival of European colonists to North America. The American chestnut, once the most valuable tree in the eastern United States, made up perhaps a quarter of the forest trees between the mid-Atlantic and Georgia. It provided lumber and food worth billions in today's dollars each year and was abundant throughout the Great Lakes region. A fungus imported on Japanese nursery stock devastated the species. In the upper Great Lakes during the 1940s and '50s, the nonnative sea lamprey destroyed a fishery worth at least $7 billion a year. And the zebra mussel, first detected in Lake St. Clair in the early 1980s, now costs the canal industry billions of dollars a year to clear fouled equipment, even as it continues to profoundly affect the Great Lakes fisheries and bird life. All too often, nonnative plants and animals become "weeds" in an ecosystem after deliberate or accidental introduction. As they overrun and outcompete natives, they destabilize and further simplify the web of life, making it easier for other invasives to establish themselves. Many of the nonnative critters and plants are of little economic value to the human components of the ecosystem. A few may even be toxic or pathogenic under some conditions.

The canal system with its hundreds of miles of feeders connects with approximately 40 percent of the state's waters, so it's hardly surprising that a number of unwanted nonnative species of aquatic plants and animals have gone canaling in order to get around. The alewife that once littered Lake Ontario's shores with millions of dead fish in the 1960s, the sea lamprey that

Zebra mussels on wall of Cayuga Seneca Canal lock. *Author collection.*

decimated high-value fisheries on Lakes Michigan and Huron and more recently the zebra mussel are among those invasives thought to have used the canal to move throughout the state. It's estimated that perhaps half of the fifty or so nonnative species of plants and animals now found in Lake Champlain got there by way of the canal system. As of this writing, two recent invaders were getting a lot of press along the canal corridor, these being the round goby and the water chestnut. Both have the potential to reshape the aquatic web of life in a huge area of North America.

The Sea Lamprey

In its native habitat, the lamprey coexists as a parasite with the large fish that it preys on. But in the simplified environment of the Great Lakes, where stocks of fish had already been affected by many years of human predation that had selected for the biggest fish and so reduced average fish sizes, its attacks became far more deadly to its smaller victims. It's generally thought that the sea lamprey entered Lake Ontario via the first Erie Canal not long after it opened, though it may have been present earlier. Sea lampreys are native to the northern Atlantic Ocean and, like salmon, run up into fresh water to spawn. The first recorded observation of a sea lamprey in the Great Lakes was in 1835 in Lake Ontario. At that time, Niagara Falls served as a natural barrier, preventing these parasitic creatures from entering the remaining Great Lakes. But by 1938, improvements to the Welland Canal (which bypasses Niagara Falls in Canada and provides a connection between Lakes Ontario and Erie) had allowed the sea lamprey to move into Lake

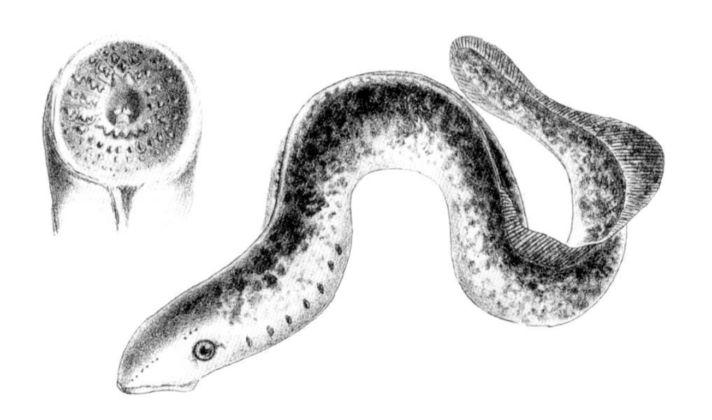

The jawless lamprey has an ancient lineage far older than most fishes and in salt water can grow to three feet. *Wikipedia.*

Erie and then throughout the entire Great Lakes system. The invaders then proceeded to devastate the commercial fishery for lake trout and whitefish. In less than twenty years, upper lakes fishery landings were just 2 percent of their peak volume of the late 1940s.

The lamprey spends most of its life buried in the gravel or sand of a clean, cool stream as a filter-feeding larva called an ammocete. The baby lamprey filters plankton and detritus and grows to about five inches before undergoing a metamorphosis that equips it with a toothy disc. In its new form, it moves downstream to open water and seeks out a fish to clamp onto. It then gnaws its way through the fish's scales and skin and draws out its blood and fluids. Big fish may tolerate the wound, but smaller ones die. After about a year of such feeding, the adult lampreys return to streams to spawn and then die.

It's been suggested that the lampreys hitched a ride to Lake Ontario on a canalboat bottom by fastening themselves to it with their discs, though no one knows for sure. And some people think they were here before the Europeans. DeWitt Clinton's journal contains this observation:

> *In Oswego and Seneca Rivers, and I think in Oneida River, considerable circular collections or piles of gravel are to be found, in the water near the shore, and sometimes on the margin of the water. Many are to be seen at very short distances, and they are evidently the work of some animal, exhibiting uniformity and design. As they appear the latter end of June, or beginning of July, when there are no freshets, and when the salmon and bass ascend, it is supposed they are erected by fish. By some they are called* bass-heaps, *and by others they are imputed to lamprey eels.*

Did Clinton observe nests made by one of our smaller native freshwater lamprey species? Or could the "considerable" collections of gravel have been nests of the sea lamprey? Some genetic evidence suggests that the sea lamprey, like the Atlantic salmon, was landlocked on Lake Ontario after the last glaciation.

Whenever they got to New York's waters, lampreys then used canals to move through the rest of the lakes. Today, the sea lamprey populations are controlled by a toxin that is applied to streams where they reside as larva. The toxin doesn't affect other fish, but some losses of insect larvae and invertebrates in treated streams do occur. As is so often the case, the arrival of an unwanted invasive causes a cascade of effects throughout the ecosystem it inhabits. And its resulting impact on native ecosystem members is often compounded by human actions.

CHESTNUTS CHOKE THE WATERWAYS

The water chestnut, *Trapa natans*, is an aquatic plant not to be confused with the edible of Chinese cooking. It first showed up in the eastern Mohawk River in the early 1900s and is thought to have been first imported to the Boston area, where it was known to have been planted in a botanical garden pond in 1877. It didn't stay there for long. Just two years later, it was going about the business of clogging up the Charles River.

Once in the canal system, it spread quickly, and today vast mats of the weed clog the edges of the Mohawk and the upper Hudson. Thousands of acres of the southern portion of Lake Champlain are now infested, and control efforts between 1982 and 2011 cost over $9 million here. Water chestnut plants form a floating rosette with a long feathery underwater attachment to the bottom muck. It thrives in nutrient-rich waters with a slow current, a description that pretty much fits everywhere on the canal. The rosette stems are tough, and it's all but impossible to get a wad of them off a boat prop unless you go overboard with a knife to hack them away. These dense floating mats of rosettes—sometimes consisting of three layers of leaves—completely block the sunlight from reaching the bottom, shading out native vegetation that fish, muskrats and aquatic birds depend on for food. Sometimes in stagnant areas with little current, once productive shallow nursery areas become anoxic as organic matter rots on the bottom. Such a lack of oxygen can devastate aquatic ecosystems, causing fish kills and leaving a turbid barren area devoid of invertebrate life that young fish depend on for food. The only way to remove a big infestation is by either mechanical weed harvest or by using a herbicide.

The rosette produces up to twenty rock-hard nut-like seeds equipped with vicious spikes. (The plant is also called water caltrop, as the seeds resemble those spiked weapons of defense that were scattered on medieval battlefields to cripple war horses.) These seeds remain viable in the mud for up to twelve years and are often spread by entanglement in the feathers of geese and other waterfowl. Because the water chestnut is so damaging, millions of dollars have been spent trying to eliminate the weed. It can cost $1,000 to clear an acre of water using a weed cutter that collects the vegetation for disposal ashore.

The costs of the water chestnut control program to protect Lake Champlain recreation and tourism activity prompted some in Vermont recently to push for closing off the Champlain Canal to prevent the entrance of another invasive, a little fish called the round goby. With their froggy eyes

Water chestnut floret. *Author collection.*

Round goby on zebra mussel bed. *Wikipedia.*

and vaguely chubby cheeks, one could say these little fish, originally from the Black Sea area, are sort of cute.

But they're also incredibly fecund and good at outcompeting small native bottom fish like the various sculpins and darters that were once important links in the Great Lakes region's food chain. It took the round goby only five years to populate the entire Great Lakes basin. The U.S. Geological Survey's invasive species website notes that female gobies can spawn at twenty-day intervals from April to September in Upstate New York waters. And the little fish can feed in complete darkness, unlike a number of potential native competitors like the mottled sculpin. They eat midge larvae, small zebra mussels, worms, scuds and other bottom life as well as fish eggs and thrive in the rocky shallows of the Great Lakes. Anglers generally despise the little egg eaters, but older larger gobies do eat zebra mussels, another invasive with a bad reputation. Besides fouling water intake pipes and boat bottoms, zebra mussels also under certain conditions concentrate deadly botulina bacteria. When gobies eat such shellfish, botulina toxin weakens or kills them. If fish-eating birds like loons or mergansers eat these gobies, a lethal dose of toxin passes on to the birds. Outbreaks of botulism have killed thousands of birds on the Great Lakes.

I first encountered the goby on Lake Ontario about fifteen years ago when crowds of juvenile fish a few inches long literally covered the bottom of an anchorage we visited in Canada. A crumb of cracker dropped in the water produced a frenzy of little two-inch fish tussling in a scrimmage for the food. A few years later, we noticed about two dozen sizable smallmouth bass under the anchorage's floating dock and far fewer little gobies on the bottom. The bass had apparently been busy thinning out the goby population. As had the local population of water snakes.

Even though cormorants, smallmouth bass and other game fish eat gobies, and even though the gobies eat zebra mussels, they are not highly regarded by many anglers or ecologists. Gobies don't get big enough to furnish human anglers with a meal, and they're extremely good at gobbling up the eggs of many of the gamefish that people do enjoy catching and eating. The various sunfishes, rock bass and large and smallmouth bass lay eggs in nests in shallows. Male fish then guard the nests from predators. However, the ubiquitous gobies make short work of any eggs left unprotected even for a brief period.

The gobies got to Lake Ontario in a ship's ballast water tank. They soon found their way into the Erie Canal perhaps by way of the Oswego River or by a connection from Lake Erie. Sampling showed their steady eastward advance until 2021, when they were at the Lake Champlain canal's doorstep. Various groups, including the Nature Conservancy, called on the state to shut down the canal to keep the little pests from entering the lake. The shutdown would have strangled recreational boater access to the lake via the canal, and at least one commercial user of the canal would have lost access for its barge loads of crushed rock bound for New York City.

A compromise for 2022 was worked out. The New York State Power Authority/Canal Corporation agreed to "double flushing" Locks C1 and C2 to reduce the likelihood of round goby getting through. They also agreed to scheduled lockings to reduce the number of daily openings, hopefully, as the news release put it, reducing the number of times round goby might "challenge" the lock system. Various agencies were sampling the canal north of the lock through the summer for signs of invading gobies as of this writing. Given the little fish's track record it seems inevitable that it will make its way into the lake, if not by way of the New York canal system then perhaps through the northern canal connected to the St. Lawrence or by way of a bait bucket, a boat's live well or some other means.

HYDRILLA THE HORROR WEED

Hydrilla, an invasive water plant that entered Upstate New York's waters from the south after originally being seeded by people dumping aquariums in Florida waters, is another invader that has used the canal to get around. Like the water chestnut, it forms dense mats that block sunlight to shade out native plants like the pondweeds and coontail. This affects waterfowl and fish that depend on these plants and associated invertebrate life. It can sometimes deplete the oxygen supply in stagnant shallows. Hydrilla showed up at the west end of the canal in Tonawanda Creek in 2012, and since it has no natural controls in our waters, it set off a number of interagency alarm bells. Hydrilla can quickly choke a waterway covering thousands of acres, making recreational boating, swimming and fishing impossible. It's been called one of the most aggressive and damaging aquatic plants known. It also has an unfortunate affinity for blue-green algae in some waters. The algae grows on the surface of the weed's leaves and can, under some conditions, produce a potent neurotoxin that, like botulism, can kill.

The algal toxin has been linked to a recently described "disease" that affects waterfowl that feed on aquatic plants. It shows up in the birds as lethal brain lesions. Studies show the neurotoxin can kill ducks, coots and geese that feed on Hydrilla. It then can be passed on up the food chain to predators of these birds like the bald eagle. A study made about fifteen years ago linked the disorder, called Avian Vascular Myelinopathy (AVM), to at least one hundred bald eagle deaths.

Because Hydrilla propagates readily from fragments, you can't cut it with a weed harvesting machine or pull it up to control it as has been done with water chestnut. It has been controlled mostly by herbicide applications, which can have unforeseen side effects on the aquatic system. It only took forty years for Hydrilla to spread from fresh waters in Florida to Maine. The Sunshine State spends $30 million a year controlling the plant from selected priority waters.

Pieces of the weed can easily lodge onto a boat trailer at a ramp approach and then be spread to new waters at the next launch. Such transport of invasive plants (and zebra mussels and other animals) from trailerable boats has given rise to a new summer job, that of the boat steward program in New York. The stewards, usually college students, are posted at strategic launch sites on busy days to check for weeds and other invasives. But it seems inevitable that Hydrilla will continue to spread.

The wondrous complexity and intricate cycling of energy and food in a biodiverse system has been compared to the human immune system. Invasive species have been likened to pathogenic bacteria or viruses that sicken humans. Sometimes in a robust biodiverse ecosystem, the invaders eventually strike a balance with the natives and coexist. But as various species of plants, animals, fungi and microbes are eliminated by human action, the resilience of the system diminishes. All too often, the invaders find themselves in an ecosystem that has been greatly simplified by habitat destruction from human activity. When nature's "immune system" has been compromised by human actions, all too often a raging infection follows, and the damage to the aquatic or terrestrial community is permanent. One unfortunate result is that plants and animals useful to humans may be eliminated. And simplified "weakened" ecosystems are unstable and inefficient and can undergo big unpredictable swings in abundances of remaining species.

It is reasonable to ask, What's next? Will the dreaded flying silver carp make an appearance in New York waters? Perhaps snakeheads from the Chesapeake or bloody red mysids will be next to push aside native life? The invasive species problem is ongoing, and it will not be eliminated anytime soon—at least as long as humans continue to move themselves and their goods and possessions around.

PART III

THE CANAL TODAY AND TOMORROW

9

KEEPING HISTORY ALIVE

Museums and More

A number of canal town historical societies have organized exhibits and collections of artifacts that preserve the history of the waterway that the present-day canal follows. The following are a sampling of a few institutions. Links to additional collections are available at the nycanals.org website maintained by the Canal Society of New York State.

The Skenesboro Museum

The canal-side museum at Whitehall occupies a concrete canal terminal building near Lock 12 and got its start in 1959, the bicentennial year for the town. Known as the Skenesboro Museum, it commemorates the town's first name when it was an eighteenth-century trading and shipbuilding center for the region. Exhibits focus on early military and maritime history using artwork, tools from farms and shipyards, ship models and a sixteen-foot diorama depicting the settlement during the Revolutionary War period.

The town was founded in 1759 by a British army officer, Phillip Skene, and was the first permanent settlement on Lake Champlain. Here, abundant hydropower supplied sawmills that along with nearby stands of virgin forest timber created a thriving shipyard. From it, workers launched a fleet of modest warships and transports during the Revolutionary War. This force under command of Benedict Arnold was defeated by the British at Valcour

Battle of Valcour Island. *Wikipedia.*

Island near the north end of Lake Champlain, but the delay of the British forces as they attempted to invade New York from Quebec is considered a key event in the Revolutionary War's ultimate outcome.

Later, the town prospered with canal-borne trade in iron, ice, eels and other merchandise while animal-powered packet boats took people from Lake Champlain to Troy, New York, in twenty-four hours.

CAMILLUS MUSEUM AND SIM'S STORE—HALFWAY THERE

The Camillus Erie Canal Park lies at the midpoint between Albany and Buffalo of the second enlargement of the Erie Canal. Founded in 1972 by Liz and David Beebe, today it is a 420-acre, seven-mile park along the former canal, now re-watered, that includes a functioning aqueduct across Nine Mile Creek. Hundreds of volunteers have worked and contributed funds over the years to make the park a historic resource of significance. The museum exhibits housed in a reconstruction of Sim's historic canal store include artifacts from all three versions of the Erie Canal along with other items of local interest. The main exhibit hall's front room is a replica of the nineteenth-century general store that once stood a half mile to the west beside a lock.

The museum has a buoy boat on display that was acquired from another museum in 2001, and the description of its work in chapter 6 was taken from the museum's research. Twelve steel-hulled boats were built in 1931 by the state to replace an equal number of old wooden tenders. Their hulls were "electric" welded, and their dimensions were twenty-eight feet (length) by seven feet (beam) with a three-foot, seven-inch draft. They were built at a cost of $1,800 apiece (approximately $32,000 in 2022). The thirty-horsepower Lathrop in the display boat is presumably original, and the original equipment list included a starting crank. Men were men back in the day, and the typical boat operator had the muscle to turn over that heavy, low-rpm engine flywheel.

Perhaps the most dramatic exhibit at the museum is the rewatered canal itself and the restored stone aqueduct that carries it across Nine Mile Creek. The enlarged version of the Erie land cut included a number of

Restored aqueduct at Camillus Museum. *Author collection.*

aqueducts, nearly all of which were demolished or fell into ruin with the coming of the current canal. Much of the stonework of the Nine Mile Creek aqueduct, built in 1842, remained in good condition, and about twenty years ago, museum volunteers began working to clear away brush and tree roots to restore it. Museum co-founder and crew leader David Beebee wrote in his book,

> *I did not realize at the onset of this restoration that many of the men did not appreciate snakes. I would show the snakes to them out of interest. Many said later that they were not fond of snakes, but did not want to hurt my feelings. On the work days I would go onto the aqueduct and hand pick the snakes and take them into the nearby woods. When the next work session arrived the same snakes were back on the aqueduct sunning themselves and I would remove them again.*

It took the Beebees and a crew of volunteers years to restore the aqueduct. Tree roots and weathering had weakened the stone support piers. After dealing with the sunbathing reptiles, the workers had to reposition the massive stones and learn about mortaring from a professional mason. They learned formulas for horizontal and vertical mortar mixes and made tools to place the mix at least three inches into the spaces between the stone blocks. Then each joint was pointed with sand and cement mix to seal the joints. The last stage was to make the aqueduct watertight by replacing the heavy timbers that made up the wooden trough that once held the canal's water. Today, crossing the aqueduct is the highlight of the museum's popular forty-minute boat ride that includes passage over the creek. It is one of a very few places in America where you can float in a boat above a creek on 180-year-old stonework.

Museum volunteers provide the public with guided tours afloat and ashore on weekends during the summer. They also run boat rides and programs for school groups. In 2021, more than two thousand fourth graders learned of life along the canal in the 1800s. Museum director Lisa Wiles notes that for some of the kids, their trip on one of the canal's several passenger vessels is their first boat ride ever. And some of the families visiting today include parents who recall their own memorable visits as kids a half century ago.

Port Byron Heritage Park

The Port Byron Canal Heritage Museum is a joint effort of the New York Canal Society and the NYS Thruway Authority and was built in 2017. Visitors can access it by either a special exit off the Thruway or from Route 31. It's open from May 1 to October 1, seven days a week. A number of artifacts are on display here, including an incredibly detailed model of the nearby Port Byron Lock 52. The model was exhibited at the 1893 Columbian Exhibition and depicts the chambers that could lock two boats simultaneously. The museum building is located on the footprint of the original Erie, as delineated by a forty-foot-wide gray swath painted across the floor. The first Erie Canal's faint depression is clearly visible just outside the building's windows. The second, enlarged Erie ran just north of the building, where visitors can walk around and through the chambers of the double lock built in 1851. It had an eleven-foot lift, making it one of the deepest locks on the enlarged Erie. The stonework of the lock walls remains as true as the day the lock passed two hundred boats during the 1860s.

The individual blocks of limestone weigh one to two tons each. They were quarried near Syracuse, transported to the site and placed using tools and

The Canal Heritage Museum at Port Byron has a special exit ramp off the Thruway. *Author collection.*

Visitors can walk through the double lock chambers a few yards from today's land-based transport of cargo. *Author collection.*

techniques that date in part back to ancient times. Beyond the locks stand three original canal-side structures: a tavern, a mule barn and a blacksmith shop. The Erie House tavern was built by an Italian immigrant named Peter (Pietro) Van Detto who also farmed a small plot of land north of the canal. Van Detto placed the two parallel bars on the tavern windows because canal-side idlers who leaned back against the windows while watching the passing traffic lock through occasionally broke the glass. His family lived in the building when it was a tavern and for many years after it closed. The family included two daughters, one of whom, Marie, recalled in the 1980s that she had watched from the window of her upstairs bedroom as the canalboats passed through the lock in the night.

The restored tavern room includes the original cash register, purchased in 1897. The museum volunteers found that the fifteen-cent key was heavily worn and surmised that perhaps this was the cost of a beer back in 1905. A restored mule barn and blacksmith shop that once stood beside the canal are now next to the Erie House. The New York Canal Society (see following section) operates the museum, owned by the Thruway Authority. The society plans future additions to the site's structures, including a restored lock tender's

The ports in the end of the Port Byron lock control the level of water in the chamber. *Author collection.*

Erie House restored tavern area with the original bar that was stored in the basement long after the business closed. *Author collection.*

building that will be placed beside the lock's entrance; a reincarnation of the Van Detto family garden, once the source of some of the tavern food served during its operation; and eventually, in front of the Erie House, a full-size wooden canalboat replica on blocks in the now dry canal bed. In the 1800s, a shipyard and dry dock operated a stone's throw from the Erie House on the other side of the canal. A boat on the hard in front of the tavern would be a fitting memorial to the vanished canal and to the tavern customers who once worked at the yard and dry dock on the boats.

When you walk through the locks, the stone walls block much of the din and roar from Thruway truck traffic that rushes past a few yards away. Stillness settles over the viewer standing within an enduring stone structure that was once as busy in its day as the nearby highway is now. The lock ruins in proximity to a fast-paced twenty-first-century transport corridor so utterly dependent on fossil fuel gives food for thought.

New York Canal Society and the Erie Canal Museum in Syracuse

The New York Canal Society was founded in 1950s and is dedicated to preserving canal history and advocating for the current waterway. It started up with the goal of saving the Syracuse weigh lock building from demolition. Today, the weigh lock is a museum supported by its own nonprofit group that also holds archives and artifacts collected by Canal Society members. Located in downtown Syracuse, where the filled-in and paved-over Erie Canal once ran, the 1850 weigh lock building is now on the National Register of Historic Places. It is the only known surviving such structure. Weigh locks were used on the original Erie to determine tolls. The canalboats were pulled into the lock chamber, the water was drained and the boat then rested on the cradle of a scale. The large mechanism worked like the balance scales we used in high school chemistry classes of my younger days. The toll per ton differed with the nature of the cargo, and the state adjusted the tolls depending on the prevailing economic conditions. There were seven weigh stations on the Erie Canal.

Today the weigh lock and its full-size replica of a canalboat contain various exhibits and host activities and programs on history related to the state's waterways. One permanent exhibit details the tools and techniques of the nineteenth-century stonecutter's art. Many splendid examples of

Syracuse weigh lock building photographed before 1915. *Library of Congress.*

1800s stonework endure along the present-day canal, often as aqueducts or as one-time lock chambers. The Erie Canal Museum also offers a variety of programs for school groups and for classroom use and supports historical research drawing on the extensive archives stored on site. Ironically, the only surviving canal weigh lock in North America was almost demolished to make way for an interstate highway. Thanks to strenuous lobbying initiated by the Junior League of Syracuse, the building was saved and the museum opened in 1962.

Each year, the New York Canal Society prints a newsletter called *Bottoming Out, Useful and Interesting Notes* for its members, and the group also organizes tours and programs on both the current canal system and past versions of the Erie. Canal Society archives include photos, maps and technical publications, some of which are available at its website. Canal Society volunteers staff the Erie Heritage Museum and Park in Port Byron from May 1 through October. Recently, the society purchased St. Johns Church, also in Port Byron, a short distance from the canal museum. The new building will provide much-needed room for archives, programming and objects that are now stored away in Syracuse at the weigh lock building and elsewhere.

BUFFALO MARITIME CENTER

The Maritime Center started as a way to get people afloat in a boat on the city's extensive and historic waterfront. Originally a program started up by a college professor named John Montegue who used wooden boat building to teach a design class, the center has since broadened to offer a range of restoration and education programs based on experiential learning. It's been said more than once that building a boat also helps build resilient young humans, and a number of programs around the country now use boatbuilding as a teaching tool for problem solving, collaborative learning and persistence. Today, schoolkids and adults alike at the Maritime Center learn about themselves as they learn to craft wooden boats under professional guidance.

Adult and young volunteers have also built several historic replicas of local significance, and a few years ago, the center began work on a full-size reproduction of an Erie Canal packet boat. The *Seneca Chief* is scheduled for launch in 2023 in time to commemorate the completion of the first Erie Canal in 1825. It is a historically accurate reproduction of the original boat used by DeWitt Clinton to travel the length of the first canal on its ceremonial opening. When complete, the seventy-three-foot, twenty-ton vessel will hopefully retrace the original route across the state and then continue to work on the canal as a floating exhibit.

On October 26, 1825, the original *Seneca Chief*, towed by four horses, departed Buffalo with Governor Clinton and other dignitaries aboard. It reached Albany on November 2 and a few days later arrived in New York City. The Wedding of the Waters ceremony was performed on November 7 off Sandy Hook, New Jersey. The boat then returned to Buffalo to complete the "wedding." Here Clinton poured a container of salt water into Lake Erie.

As of this writing, the professional staff and volunteers were well underway on the replica's construction. Other historic replicas in the center's fleet include a traditional Lake Erie fishing shallop dubbed the *Buffalo Wailer* and the *Erie Traveler*, a fifty-two-foot replica of a Durham boat for the Lockport Flight exhibit housed in the old lock powerhouse. Durham boats were widely used on the Mohawk and other pre-canal waterways in colonial times.

These are just a few of the museums along the canal today. Check the Canal Society's website for listings of others along the route of the current waterway and that of the previous canals.

WHY MUSEUMS MATTER

Much of history is preserved in prose and static illustrations. A museum, however, with its physical artifacts and human activity, presents history in a way that involves the whole mind and the body. Such presentations can even change the course of a life. Seeing artifacts once used with skill and sweat and sometimes considerable muscle to make the waterway function prompts all sorts of thoughts and associations on the viewer's part. And knowledgeable guides, like those who staff these canal museums, make a visit special. Revelations gleaned from a museum can and do shape an individual's world view.

The canal societies, museums and other groups with a maritime focus have dedicated, passionate volunteers and staff and are doorways to the past, lest we forget. Museums spark curiosity. They arouse admiration for the ingenuity and craftsmanship of our ancestors. And they entertain. Today, people delight in their canal-side experiences at Camillus and marvel at the changes in travel and transport seen so vividly at Port Byron's museum

"Traveling with the softness of a dream." *From* Harper's *Magazine, 1873.*

beside the New York State Thruway, one of the busiest toll roads in the United States. As I write this, volunteers and skilled craftsmen are taking pride at working together to shape the timbers of a splendid wooden vessel that will soon reenact the *Seneca Chief*'s transit of the canal. And museum scholarship in Syracuse and elsewhere continues to provide insights as to how we got to our present. We cannot understand or effectively deal with today's society and its relationships, conflicts and problems without an understanding of historical facts. Museum workers dig up those facts from old letters, account books, publications and objects. Did the original Erie Canal excavation use slave labor? Why would First Nation elders take a dim view of a recent reenactment proposed on the canal? How is the current canal influencing ecology and society along its length? History helps us find our way into all that future that lies before us.

10

TRAVEL ON THE NEW YORK STATE CANAL SYSTEM TODAY

The Oswego Canal and Port

The original Oswego Canal connected the Great Lakes system to the Erie in 1828 by using the river with its several sets of rapids and waterfalls. It was completed in 1828 after three years of work and followed the river north to Phoenix. After that, a line cut allowed descent around the falls and rapids via a series of locks to Oswego's harbor. The connection was an immediate commercial success, with the salt trade a mainstay early on. By 1855, vast amounts of flour from the water-powered mills of Oswego along with western corn, barley and oats were being transported to New York City on canalboats. The enlarged nineteenth-century version of the waterway also made Oswego one of the biggest lumber handling ports in the nation during the 1870s. Millions of board feet of lumber from the pine forests of the upper Great Lakes region came to the harbor aboard schooners and steamers after the Civil War. It was then transported downstate on canalboats to help build New York City's factories, apartment buildings and warehouses.

Once, Oswego rivaled Buffalo and Chicago in importance on the Great Lakes, and one of the most important maritime innovations of all time, that of the screw propeller, was first used on a commercial vessel by an Oswego businessman named James Van Cleve. (For more on his life and times, check out the chapter on his work in *Maritime Tales of Lake Ontario*.) Today, Oswego remains an active port, but very little of its commercial traffic uses the canal now.

Port of Oswego lighthouse built in 1934 on the end of a two-thousand-foot stone breakwater. *Author collection.*

The million-bushel grain elevator that once stood on the west side of the river where it loaded canal barges with twenty thousand bushels of wheat in an hour has vanished. It was demolished in 1999, as highway construction and the St. Lawrence Seaway had made it obsolete. Powdered cement still comes into the port, where ships transfer the product to silos after loading their cargo from Canadian quarries on the lake's north shore. Its subsequent movement and that of the grain stored in the port's new storage bin depends on truck and rail traffic rather than canal transport.

As recently as the 1990s, the Oswego section of the canal system saw movement of cement aboard the *Day Peckinpaugh,* barge loads of corn for the ethanol plant in Fulton and occasional other cargoes. Today, the asphalt, fuel oil for the steam power plant on the city's shore and cement that arrive by ship no longer travel from Oswego on the canal.

A few hours' travel south from Oswego takes the canal navigator to the falls and rapids of Fulton. Here abundant hydropower, still used to generate electricity, once ran a number of industries. Remnants of the old factory building walls and foundations still stand beside the waterway, as do the structures of Huhtumäki, a Finnish manufacturer of paper food packaging. It occupies an area on the canal's east side that was formerly owned by Sealright Corporation, known for the invention of paper ice cream and milk cartons. This site may be the oldest continuously operating corporate concern on the canal, going back to 1883, when the Oswego Falls Pulp and Paper Company first used the river's hydropower and wood that was then delivered by canalboat. Today, the factory continues to operate beside the canal thanks to growing interest in sustainable food packaging. Some of the company's products use a substantial percentage of recycled paper fiber. A study (funded in part by Huhtumäki) states that a compostable paper cup has a far lower carbon footprint than a reusable plastic cup that must be washed after each use. And high-quality fiber used in a paper cup can be recycled seven times.

THREE RIVERS TO BUFFALO AND THE LONG LEVEL

After traveling the thirty-eight-mile length of the Oswego Canal, the vessel navigator comes to a fork in the road called the Three Rivers Junction. Here you can make a right onto the Seneca River and head for Buffalo, or you can part ways with the Oswego River to steer east down the Oneida River, ultimately to tidewater at Troy. Either way takes you along the route of the original Erie. To the west, once past Lyons, the current canal leaves the Seneca and Clyde Rivers to become a man-made ditch following the original Erie's footprint. This section is the sixty-six-mile "long level" that has no locks once past Pittsford and Rochester, though a slight grade moves Lake Erie's waters eastward. It cuts right through a string of towns and small cities that were built along the original canal, and here the recreational boat captain is faced with a new adventure—the lift bridge. These are truly low bridges, often with only six to eight feet of clearance above the water. Most open on receiving notice from the boat skipper, usually via cellphone these days.

Past Lyons, the original towpath closely hugs the waterway for most of the trip to Buffalo and is popular with pedestrians. Around Rochester, its paved surface makes the car-free trail a cyclist's delight. Ten speeds, electrified fat tire bikes, hand cycles, recumbents and trikes of various configurations all speed along beside the placid water. The Canal Corporation and other nonprofit organizations organize and promote group rides across the state on the towpath for cyclists each summer. The flat terrain and mostly off-road trail attract hundreds of pedalers of varying levels of fitness and ambition. Group rides often include a sag wagon and an itinerary that allows access to a hot shower and a motel bed each night.

The stretch of waterway around Rochester posed unique challenges to the builders of the Barge Canal and was the last section completed. The former canals had cut right through the city via an aqueduct over the Genesee River. But the new, larger channel had to be routed in a considerable loop south of the city partly because of the cost of acquiring a wider right-of-way through downtown. A two-hundred-plus-foot-high ridge of glacially deposited gravel and rock blocked the route immediately south of the former canal and so had to be avoided. But the canal plan then had to cut through Genesee Valley Park, and the residents did not want their park fragmented and defiled by its channel or unsightly high levees. The protective banks were graded, and a series of handsome foot bridges solved the park access and aesthetics issue. A guard gate on either side of the Genesee protected the canal from extreme changes in the river's flow.

On the other side of Rochester and again at the approaches to Lockport, the canal builders had to make formidable rock cuts.

The task just west of the city was compared at the time to the deep cut of the Panama Canal at Culebra. The contractors used a variety of equipment to drill, blast and break rock. Whitford describes a piece of equipment known as the rock breaker as follows: "The shattering (of rock) was accomplished by dropping from a considerable height a cylindrical hammer, 26 feet long and weighing between 15 to 16 tons. At the lower end of the hammer a removable section, in shape a conical point, could be renewed when worn out." This outfit floated aboard a scow held in position and moved by means of wire cables anchored on shore. Then the stony debris had to be mucked out by excavators.

At the Rochester cut, another device, the Bridge Excavator, aka the grab machine, spanned the width of the channel traveling along on rails as the cut proceeded. It lowered a huge steel grab bucket to seize the rocky spoil. Then the bucket was raised seventy feet and moved to one end of the bridge to be dumped on shore some distance away from the channel.

Today, the old commercial canal terminal at Rochester reached via the Genesee River has been converted into a recreational oasis and waterfront residential area called Corn Hill Landing. Here visitors find various amenities, including dockage for boaters and plenty of restaurants within walking distance.

After a day or two of travel through the flat lake plain countryside of farms and towns and villages—many with *port* in their names—you arrive at Lockport, where the canal climbs the Niagara Escarpment using two high lift locks that adjoin each other. They replaced two flights of five locks side

S.T. Crapo built in 1927 by Great Lakes Engineering Works, seen here by Oswego cement silos in 1934 (note the grain ship unloading behind it). *Richard Palmer collection.*

Dikes along the canal protect the park from floods, and bridges "of artistic design" allow pedestrian crossings. *Author collection.*

At Lockport, the rock of the Niagara escarpment had to be excavated for the flight of locks for the various versions of the canal. *Author collection.*

by side, one of which remains in view today. The gorge below the lock was carved by an ice age Niagara-like outflow from Lake Erie. Today a guard gate and lock doors hold back the lake.

The New York Canal today terminates at the Niagara River, and contemporary mariners must navigate the mighty outflow of four Great Lakes for a few miles before reaching the entrance of the Black Rock Canal, a federal waterway, which bypasses the swift currents in this section of the river. Buffalo has paved over the historic Erie Canal, and little trace of it remains except for a tiny inlet near the Naval Park and at the Erie Basin Marina, where once canalboats rafted up waiting to collect grain cargoes. But it was the Erie that for a few years made Buffalo into the greatest grain port in the world. Today, most of the massive silos now stand silent and empty along its waterfront, though they remain as monumental reminders of that day. Some grain does still come into the city by ship and by rail, and General Mills still maintains a silo block on the harbor.

Much of the activity on the waterfront is recreational, and efforts are underway to reestablish the city's historic relationship with water. Colored light shows play on the concrete silos of the harbor's Terminal Block, and music festivals and concerts exploit the amazing acoustics of the man-made

The Peace Bridge connects Canada and the United States across the Niagara River, seen in the background in this view from Black Rock Canal's quiet waters. *Author collection.*

The Great Northern Block in 1985. Demolition of this block of silos was underway in September. *Wikipedia.*

Most of the city's silos are empty and unused today. *Author collection.*

canyons of the waterfront. Festivals, paddling and walking tours and boat rentals all enliven the summer waters of the harbor and ship channels. Buffalo also has a hearty craft beer industry, and one of the silo groups was recently repurposed to house a brewery and painted to form a Labatt's six-pack in recognition of this pillar of the Queen City's tourism.

Eastern Section Oneida Lake, Mohawk Valley to Tidewater

If you make a left turn at the Three Rivers Junction, your destination may be either New York City via the Hudson or Montreal via the Champlain Canal. Because the canal between Oswego and Albany has higher bridge clearances, it is sometimes used by larger private or commercial vessels on their way to or from the Great Lakes, and you never know what you might encounter as you travel along. The occasional multimillion-dollar mega yacht, replicas of the *Pinta* and *Niña* built in Brazil, assorted versions of nineteenth-century "pirate" ships and small tall ships (with their sailing rigs on deck), a Viking ship replica, and a Polynesian sailing canoe (the *Hokule'a*), which sailed from Hawaii to the East Coast and visited the Great Lakes via the canal, have all traveled this stretch of water in recent years. They follow in the wake of a motley fleet of vessels that in the past carried cargoes of ideas rather than lumber, coal or oil. Theater boats, traveling side shows, missionary vessels and even a barge with the front portion of a preserved whale once plied the nineteenth-century versions of the canal.

One of the oddest ships that I have ever encountered on the canal was the steel-hulled sailing barge *Amara Zee*. It's been called the most interesting ship in the world. When I saw it in 1999, I would have said it was perhaps the most incomprehensible. Loosely based on the design of a traditional Thames River sailing barge, its steel hull was built in Kingston on Lake Ontario over a four-year period for the Caravan Stage Company and was launched in 1997. *Amara Zee* then went off on tour traveling the Intracoastal Waterway of the East Coast. A *New York Times* reporter described its presentations as "opera that might be described as Cirque du Soleil meets Occupy Wall Street."

We encountered it in Waterford when the crew was presenting *Shakespeare's Dog*, a play described in an online review as "absurd, insightful and occasionally confusing." I would go with the last. I was totally confused. But the acrobatics of the actors in the ship's rigging were entertaining. I saw the barge again in 2015 in Oswego when it used the canal to return to its natal waters of Lake Ontario. The performance was quite political in nature, and about half the audience walked out during intermission. But controversy has been no stranger to the nature of theater. Drama has pushed the boundaries of social commentary for a long time. Many historians have written about the effect of the original Erie Canal on society. It's been said the flow of progressive ideas and religious fervor that came with canal travel and its commerce changed the nation profoundly by helping bring about movements for women's rights and the abolition of slavery. So the *Amara Zee* was following a hallowed tradition in its inland odyssey.

After a day aboard a slow boat heading east from Three Rivers, the modern-day canaller is confronted with the wide waters of Oneida Lake. It seems almost oceanic after chugging along the canal for a day or two. The hazards and rocks on either side of its well-marked channel don't invite exploration by a through traveler, but the shallow, warm lake has a long history of productive fisheries. Its twenty mile east–west fetch is also capable of kicking up a nasty steep chop, and canallers with small, slow sailboats carrying their spars on deck treat it with respect.

When the Barge Canal first opened, a number of old wooden canalboats leftover from the towpath era were still afloat, and some of them got into trouble now and then when a late fall gale kicked up on Oneida Lake. For a time, the state kept two sturdy steam-powered rescue tugs on duty to assist distressed tows. One storm saw five wooden barges break loose to go ashore. The tug *National* hauled them off one by one and got them and their tug safely into port. Verona Beach on the lake's eastern shoreline was known as the lake's graveyard for the more than a dozen wooden barges that were caught by northwest gales and beached here. Another fall disaster on the lake occurred in 1936, as detailed in Michelle McFee's book *A Long Haul*. That year, a sudden, hard freeze caught dozens of tugs and barges on their last trips of the season in mid-November. A news story recorded that the boats on Oneida Lake faced a cold and possibly hungry winter. With food and fuel provisions dwindling, seventy men aboard 5 stranded motor ships were left hoping for a wind shift to blow the ice away. It took a one-thousand-ton tanker and two state tugs, the *Seneca* and the *Syracuse*, to break a channel in the twenty-inch ice for the boats to get off the lake and safe into winter quarters in Rome. Of 385 barges and tugs caught by the freeze that year, about two-thirds spent the winter in protected sections of the canal.

After Oneida Lake, the canal summits at Rome. It then passes through perhaps the most scenic stretch anywhere on the main east–west line as you enter the Mohawk drainage. East of Oneida Lake, the canal passes along a low-lying rural stretch of countryside that once was the route of the early travelers using Wood Creek. We spent a quiet early summer night at remote Lock 22 some years ago. Here, for the first time, I heard the canal's singing fish. Though not as melodic as a robin or a wood thrush, the freshwater drum (aka sheepshead) is incessant—and quite loud, especially if heard through the hull of a boat when you are below decks. They sound a bit like distant throbbing engines. When I checked the navigational chart, I saw we had tied up near where Drum Creek entered the canal.

Today's waterway bypasses Rome. But at the time of the original Erie, it cut right through the city. At this point, the waters on each side of the summit flow away from Rome. To make up for this loss, the waters of the Hinkley and Delta reservoirs enter the canal bed. Feeders from the old Chittenango canal and various reservoirs to the south also flow in to supply the water that drains east and west from here.

The Mohawk Valley was a key strategic area during the Revolutionary War thanks to its fertile wheat fields. Had the British gained control of it, they would have cut the rebel army off from a major source of food. The Battle of Oriskany, located a few miles east of Rome, was one of the bloodier battles of the Revolutionary War. Here, a force of Mohawk, Seneca and Loyalist fighters ambushed General Nicholas Herkimer's revolutionaries plus a small force of Oneidas who had sided with the Americans. They

Tug *Syracuse*, built to be ice-worthy in 1934. It is seventy-four feet in length and has a 575-horsepower diesel engine. *Author collection.*

It's believed that the drum, sometimes called croakers, which "sings" in early summer, does so to attract a mate. *Wikipedia.*

killed, wounded or captured more than half of the rebels. But when word of the attack reached Fort Stanwix at Rome, the Americans were able to turn the tide of a British advance from the north, a major strategic delay that helped bring about the support of the French and the eventual American independence.

A few miles east of the battlefield, the traveler passes Frankfort, where the Mohawk River joins the land cut. Here we had our own battle with the bottom mud during a canal trip in 2020, when we ventured up the river in search of a town dock to spend the night. We got well and truly stuck, and as we sat motionless in a light rain, we marveled—who but an utter idiot could run aground in the canal? It took several muddy, wet hours of kedging, heaving and hauling to get our boat afloat again.

Actually, running aground in the canal can and does happen, and to prevent it, the state invests considerable effort and money in dredging the portions of it that follow river channels. Both floating dredges and shore-based excavators maintain channel depths, removing 280,000 cubic yards of material on average each year. Hydraulic dredges pump sediments through pipelines to upland disposal areas, where solids are allowed to settle out before the water drains back into the canal. Coarser sediments are excavated by clam shell dredges or shore-based excavators and loaded onto barges for transport to disposal sites. Recently, the Canal Corporation has done more dry dredging with conventional land-based equipment during the off-season when water levels are drawn down. Winter dredging has considerably less biological impact on the canal's fish and other organisms.

And this stretch of the canal has plenty of fish in it. It's common to see anglers fishing from boats below the movable dams or from the approach walls of the locks, while both eagles and ospreys are also seen perched in trees in wooded areas along shore. Besides the aforementioned freshwater drum, the canal provides anglers with northern pike, smallmouth bass, walleye and other game fish. And carp. The common carp, some of heroic size, flourishes throughout the waterway. There are carp derbies and contests and guided fishing for carp along the canal's length. The Wild Carp Club of Central New York promotes the quest for what some consider to be the world's greatest sport fish. Carp are big and brawny and do put up a terrific fight when hooked, but after the battle, most anglers practice catch and release. Few seem excited about carp for dinner. The record carp caught in New York was a fifty-pounder, but those who seek these hardy fish know there are bigger ones out there somewhere lurking under the surface in the backwaters of the canal.

Carp from the Cayuga Seneca Canal are blocked by a flow control structure from the warm weedy waters of the Montezuma swamp where they seek spawning areas in the spring. *Author collection.*

Little Falls, home to Lock 17, the highest lift on the system, is an old manufacturing town and a good place to observe canal-side geology. This is a scenic stretch of canal thanks to the force of the ice age outflow of water from the Great Lakes. That outflow cut its way through the rocks that today make up the walls of the gorge that rise above the river. It also gave rise to a unique geological feature on Moss Island beside the lock. Here a series of round "potholes" some many feet across and thirty feet deep were formed by the mighty cataract of water that flowed forth from the lakes about fifteen thousand years ago. They may have been carved out of bedrock by grinding swirling stones at the waterfall base, though no one knows for sure how they were made. In more recent times, the forty-foot drop in the river powered a variety of nineteenth-century mills and factories. The canal also served paper mills and other manufacturing concerns. Today, Little Falls maintains a splendid city landing to welcome canal travelers with all the amenities required for long-distance boaters, including a laundry room with washer and dryer facilities for the muddy boater, something our crew appreciated greatly back in 2020 after our grounding.

A short distance to the east of Little Falls near Fonda, the river passes another pinch point in the highlands. In this narrow gap, railroad, highway and canal carry traffic within a stone's throw of one another. The profound contrast of swift, noisy autos and trucks and trains with the slow-moving boat is a marvel to the modern-day canaller. As we rounded the bend and gawked at the Thruway shoulder just above us, a trucker waved from his cab. A bit farther on, our vessel got a short toot from a passing locomotive engineer. Pushing on east takes you past the one-time industrial heart of the upstate where Ilion, Canajoharie and Amsterdam once produced carpets, buttons, Remington rifles, typewriters and lots of baby food. Here, while locking through, we viewed the flood-damaged remains of Guy Park's manor house, a reminder of the Mohawk River's sometimes untamed power.

At Waterford, the canal steps down into the Hudson River, where the traveler meets another junction. You can choose to continue canaling by taking a north heading to Lake Champlain, or as most long-distance travelers do these days, you may swing south down the Hudson to salt water.

Cayuga Seneca Canal

If a helmsman takes a westward heading at Three Rivers for a freshwater cruise to Lake Erie, he or she soon encounters a turnoff for the Cayuga Seneca Canal. When the canal enlargement was first proposed, this twenty-mile connection between the two biggest Finger Lakes and the Erie was to be merely a feeder. However, large salt deposits on both lakes and vigorous lobbying by local businesses and politicians convinced the state to enlarge the existing canals and cut a channel four and a half miles south along the edge of the Montezuma Wildlife Refuge to Cayuga Lake. The channel also goes west to Seneca Falls and Waterloo to connect directly to Seneca Lake.

Excavating the enlarged channel for the Cayuga Lake outlet along the edge of Montezuma's Great Swamp was a distinct challenge. The deepened canal lowered the marsh's water levels several feet, and when the Federal Wildlife Refuge was established here a few years later, large areas of wetland had to be rewatered by an extensive system of dikes and embankments. Today, the refuge includes about ten thousand acres of seasonally flooded lands and wetland, about half the size of the original marsh. It remains a place of wide skies, great flocks of migrating waterfowl in spring and fall and seemingly endless cattail and reed beds.

Once the wetland supported a considerable cottage industry of fur trapping, especially for muskrats. The fur trade was an important supplement

"Ray" the heron at May's Point Lock waiting for the discharge turbulence to turn up a fish. Herons, ospreys, eagles and cormorants abound along the canal where it skirts the National Refuge. *Author collection.*

to meager Depression-era farm incomes in the 1930s. An old-timer told me that the skinned carcasses were placed on trays to be picked up by an entrepreneur who sold them to meat markets down on the Chesapeake. (In the 1970s, on the Nanticoke River I saw scrambled eggs with muskrat brains on one diner's spring breakfast menu, and skinned 'rats were lined up in a grocery meat display counter. I was told the taste was similar to rabbit.)

A few years before the enlarged canal was built, a short-lived experiment in cardboard production took place on the edge of the Great Swamp as detailed by Mike Riley in "The Cattail Company" on the NY Almanac website. The Montezuma Fiber Company set up a factory to produce boxes made from cattail flag pulp. The flag was cut into small pieces and then cooked by steam in a giant vat. From there, the softened flag was piped into a shaper that flattened the mix into board-like sheets. Steam-heated rollers continued to flatten and dry the pulp, and the end result was more board than paper. It was used to make boxes and lining for shipping crates. The Fiber Company even mixed it with leather scraps to produce shoe soles. However, the business didn't last long. The new canal drained and shrank the wetlands, and new technology made it possible to produce inexpensive cardboard from wood pulp. After just ten years, the factory went dark as the company declared bankruptcy. But to this day, a few people still harvest flag from the marshes for use in making rush seat furniture and for woven mats and baskets.

The canal enlargement eliminated the "falls" from Seneca Falls and flooded a substantial portion of the city and its hydropower-dependent mills. Today, the reservoir where the one-time rapids ran about a mile through the area then known as the Flats is called Van Cleef Lake. The state demolished over one hundred businesses and about fifty houses to clear the area before flooding it. Today, the lake supplies water to the forty-nine-foot lift of the combined tandem locks here as well as to the hydropower station beside them.

Seneca Falls today has several museums detailing the industry of the area that once called itself the pump capital of the world. To this day, the city is headquarters to ITT Gould's Pumps, an international corporation that was established in 1848. It also has the distinction of being the location of the 1848 convention that drew up the Declaration of Sentiments under the leadership of Elizabeth Cady Stanton, then a resident of the city, and several other women who launched a nearly century-long struggle for women's right to vote. The iconic canal-side stone knitting mill that once produced socks for the Apollo astronauts is now being remodeled to house exhibits and programs related to the history of the suffrage movement.

This effort followed other earlier political actions for temperance and the abolition of slavery. It drew energy in part from the change nourished throughout the region by trade and travel on the original Erie Canal. New ideas, businesses and relationships fostered by the canal gave rise to social and religious movements of national significance. That same progressive

Seneca Falls knitting mill, built in 1844, produced woolen goods until 1999. *Author collection.*

notion that government could promote general social good fueled the first canal's construction. Ultimately, it also made possible its last enlargement.

Seneca Falls has created an attractive landing for boaters to tie up to. One of the buildings beside it houses the Seneca Museum of Waterways and Industry, with excellent exhibits detailing the industrial and social history of the city. Here products of the industrial powerhouse that was Seneca Falls are displayed—flat irons, pedal-powered scroll saws, fancy cast parlor stoves, knitting machines and, of course, pumps. After passing through Waterloo and another lock, the Cayuga Seneca Canal terminates at the north end of Seneca Lake.

CHAMPLAIN CANAL

The Champlain Canal runs about sixty miles north from Troy, much of it following the Hudson River. At Fort Edwards, it leaves the river to become a line cut that ultimately connects with a long, narrow extension of Lake

Champlain called South Bay 95.6 feet above sea level. At its summit, 140 feet above sea level, the Glens Falls feeder enters it, bringing water in from the Hudson. It then descends to Whitehall, where it ends. Its total elevation change is accomplished by eleven locks. This was the first segment of the original Erie to be completed, and it follows a historic water-based route of great strategic importance in the first decades of our nation's existence.

After the last ice age glaciers gouged out a pre-existing fault depression to form Lake Champlain, that waterway provided a vital north–south highway to the early European colonizing powers of France and Britain. Various battles and skirmishes dating back to Samuel De Champlain's fateful military assault on an Iroquois village in 1609 mark the history of this region. Fort Ticonderoga, a few miles north of the canal's termination, saw key battles during the Seven Years' War and the Revolutionary War. The Battle of Saratoga, considered the turning point of the American Revolution, took place a few miles from the present-day canal. At Schuylerville just south of Lock C-5, the Saratoga Battlefield Monument recalls this key struggle. Again during the War of 1812, the region saw naval action and ample bloodshed.

Tug with barge load of crushed stone on Champlain Canal. *Rob Goldman.*

A few miles to the north, the canal and the river part ways. Here the village of Fort Edward has a museum complex of several buildings recalling Revolutionary War action. The canal terminates at Whitehall, said to be the birthplace of the American navy. A museum located in a former 1917 canal terminal contains displays on the region's military and industrial history. In 1775, the British forts at Ticonderoga and Crown Point at the south end of Lake Champlain were captured by the rebels, and a small merchant vessel built at Whitehall was seized, renamed the USS *Liberty* and set sail to successfully assault a British shipyard near the lake's north end. This gave the Americans control of the entire crucial travel route. Several additional small warships were then built at Whitehall's shipyard to transport troops for the invasion of Canada. The invasion, however, did not succeed.

Most of the traffic on the Champlain Canal today is recreational, but since 2016, the NYS Marine Highway Transportation Company has moved crushed stone for road paving through its length and down the Hudson River to New York City.

Today this serene land of ancient mountains and attractive villages and towns has some of the most scenic landscapes of the entire canal system. Through this canal, the traveler can also make an international connection by traveling the length of Lake Champlain to reach the Chambly Canal in Quebec. From there, a boat can reach the St. Lawrence River, which descends to the sea.

11

NEW IDEAS ALONG THE OLD CANAL

A few months before this was written, an oversized load made its way to the coast via the canal. The tugboat *Edna A.*, owned by the New York State Marine Highway Transportation Company of Troy, headed east along the main canal corridor pushing a barge loaded with a nuclear submarine condenser built by a Batavia-based company. The 100-ton condenser was trucked a short distance by road to the canal terminal and transferred to the barge. It traveled to Albany and then down the Hudson to New York Harbor, where the cargo was transferred to another oceangoing barge for the last leg of the trip to the New London, Connecticut sub base. Other specialty cargoes too big to move by truck also still travel the canal on barges. They include boilers, transformers, turbines and, in 2017, a dozen two-thousand-barrel beer tanks that went from Albany to a Rochester brewery aboard four barges. That year, 415,000 tons of freight moved on the canal, much of it being stone from quarries near Whitehall on the Champlain Canal that was barged to New York City for use on roads. There have been periodic proposals to move NYC garbage on the canal to upstate landfills, and other efforts have been made to move recycled material on the waterway. But in recent years, recreational traffic has dwindled, and the NY Power Authority has shortened the navigation season as a cost-cutting measure. A number of people have asked why not simply abandon the canal? Why continue to spend $80 to $100 million a year mostly to support a few thousand recreational boat trips? And why risk the introduction of more invasive species to our waterways? Let's save money and shut the waterway down. Far more people use the towpath than navigate its waters on long-distance cruises.

Cities and towns along the five-hundred-mile system have spent considerable energy and money on canal-side parks, facilities for traveling boaters and public access. A lot of the energy has come from civic-minded volunteers. These public landings beside a functional waterway with their docks and green spaces add interest and vitality to the municipal waterfront. People fish, bring the kids down to the water, birdwatch or simply sit on a park bench and watch the traffic go by. Waterfront restaurants and cafés thrive during the summer season, and canal-centered festivals and activities generate revenue for various small businesses ranging from marinas to gift shops. And non-boaters enjoy relaxing boat rides aboard various sightseeing tours that depend on public landings for their passengers to access the vessel. To this writer, abandoning the canal as a navigable waterway seems like a betrayal of these many public and private for-profit endeavors.

People have been reimagining the New York Canal system for more than twenty years. One of the more audacious efforts was that of Dr. David Borton, a retired engineering professor. In 2015, he made a carbon-neutral delivery to a recycling mill of four tons of cardboard. He did it by traveling the entire east–west corridor of the waterway on a solar-powered boat of his own design. *Solar Sal* was forty feet long and was built in part by volunteers and students at an Albany-area high school. Its battery bank was replenished by panels mounted on the ship's upper deck and allowed travel on cloudy days at reduced speed. *Solar Sal* had enough battery capacity to run all night, and it made the 750-mile round trip using only sunshine.

This "proof of concept" voyage led to the launch of several more solar-powered boats, including *Solaris*, a forty-four-foot Coast Guard–certified passenger vessel that now operates on the Hudson River. It has a fifty-mile range with no solar input while lacking the noise and vibration of diesel power. Borton and his co-workers at Sustainable Energy Systems recently built another solar-powered boat for use on Puget Sound and also offer several designs for smaller solar-powered boats for recreational use.

Electric boats that use the grid rather than the sun for recharging batteries have been around for a long time. Way back in 1893 at the Columbian Exhibition in Chicago, rides on the fairground ponds and lagoons aboard silent running electric launches were hugely popular. Advances in battery technology have brought new attention to this old idea since the 1990s, and a variety of battery-operated outboard and inboard motors are increasingly being put into service. Recently, New York State announced an incentive

Solar Sal traveled hundreds of miles powered only by sunlight on the canal with cargo. *Author collection.*

program to support the conversion of commercial charter and passenger-carrying tour boats from diesel to electric propulsion. Details including charging infrastructure enhancement are still being worked out, but one charter company, Erie Canal Adventures, has replaced a boat engine with an electric motor and is now testing the outfit on the canal. The simplicity of operation and lack of fumes and noise are attractions for charter boats that are operated by the clients. Adding solar panels to electrified vessels would reduce the load on the grid at charging stations and increase the range of the boat. During a canal trip in 2018, I encountered a hybrid diesel electric houseboat named *Dragonfly* that had traveled the canal with diesel engine, electric power and solar panels to charge the battery bank. The skipper estimated that his set up cut his fossil fuel use by 40 percent.

In 2018, the towing firm New York State Marine Highway moved 500,000 tons of crushed stone for asphalt paving using the canal. According to an engineering report done by Texas A&M, inland towing averages 647 ton miles per gallon of fuel. That compares with 145 ton miles per truck, and a single barge may carry 100 truckloads. When fuel costs spike (as they inevitably will in the future), canals and the lakes again could become a cost-effective way to ship between the eastern and midwestern United States. Short-haul bulk and heavy oversized cargo are carried on European canals and rivers, contributing about $80 billion a year to economies there. Shipping on the Danube was a lifeline to the Ukraine for moving barge loads of grain during the early months of the war with Russia in 2022. Containerized freight has also been shipped on barges in Europe, and using such transport here could reduce traffic congestion and greenhouse gas emissions. By offering an alternative to the ubiquitous long-haul truck, it could create a more robust and resilient transportation system within the state.

One problem in Upstate New York is that of terminal access, where cargo can be transferred from barge to land-based transport. And speedy transport is presently not an option on the canal. Still, interest in sustainable energy use is on the increase and with good reason. As climate change accelerates, efficient low carbon–emission transport is imperative to our future survival. Dr. Borton of Sustainable Energy Systems points out that just a few generations ago, all shipping was solar powered. Four-legged "hayburners" fueled by green plants towed the canalboats, while on open water, wind-powered schooners and square riggers carried the world's maritime trade. Solar panel efficiency and technology have advanced rapidly, and there are now prototype electric autos equipped with solar panels and batteries being tested that can travel forty miles on a day's worth of sunshine power. The Upstate New York canal system is inherently seasonal, but during summer's long days, we enjoy plenty of solar input.

Unfortunately, for those with interest in boat travel, many of the reimagining proposals for the canal that have been funded are based on real estate values and towpath use. It's been suggested that sections of the canal be drained to prevent the spread of invasive weeds and animals. If this were done, people might be able to paddle portions of the future waterway with small boats, but the through traveler would be excluded without the locks and dredging that maintain the waterway today. Today's canal system also plays an important role in flood control, and some aspects of its various control structures would have to be maintained even if navigation were shut down.

Last Word

"Is the canal still functional?" a young sailing enthusiast asked me recently. He seemed a bit bemused by our decision to charter a houseboat for a canal holiday. But muddy water glitters like silver in the sunlight, and the canal, viewed by many upstate residents as a dull, stagnant ditch, still

Boating and feeling fine on the Mohawk River, but for how much longer? *Author collection.*

surprises and delights the traveler. These days, thanks to the Clean Water Act of 1972 that funded municipal sewage plants across the land, the canal is filled with fish and fish-eating wildlife. No bird is more emblematic of it, I think, than the great blue heron so often seen standing sentinel along its shoreline.

Rivers, often viewed as a metaphor for life, twist and turn, run slow or bound and leap frothing over rocks and rapids. The predictable canal lies long, straight and sometimes quite narrow before you. It's not unlike a secondary gravel road out on the linear landscape of the western prairies with their straight horizons. Yet the canal still attracts recreational travelers in search of a novel experience. During a recent sojourn with the rented houseboat, we saw a woman paddling along in a kayak, its decks piled high with lashed bundles of gear. Clearly, she was on a long journey. A lock tender told me he had passed an elderly man and his rowboat through his lock who had rowed from Lake Erie and was bound for salt water. I met a young student once aboard a homemade raft propelled by a small outboard. He had planned and dreamed about his trip for years. He told me as he prepared to get underway for Oswego and the canal that he didn't hope for good fortune, he depended on it.

It's hard to explain the appeal of canaling. Aboard a boat underway, the shoreline scrolls by at a leisurely pace that allows one to comprehend, ponder and discuss features seen. Look at that heron, someone says (it is perhaps the tenth such sighting in the last half hour). Or you wonder about an abandoned little cabin standing in a wooded clearing on shore and picture yourself sitting on its porch during a summer evening, an easy cast of a fishing line from the water. There's time to see an osprey snatch a fish dinner and fly off or to marvel at dozens of densely packed boils of minnows circling in tight swarms at the surface around you. Grand vistas are not the rule along many stretches of the waterway. Rather, you peer into the dense growth of vine-clad trees and shrubbery along the banks wondering what lives in the deep, dark green forest. Some of the more rural sections of the waterway recall images of the Amazon or the Congo as seen in nature documentaries. Surely there will be a crocodile or perhaps a hippo around the next curve. Instead, a pair of derelict oil storage tanks loom dark and rusted through the brush. Sometimes we went for several hours without seeing another boat. But we saw anglers, towpath cyclists and walkers along the canal. And our chartered houseboat attracted visits several times from curious children when we tied up at a village landing. They wondered what it was like to live afloat.

Most of the privately owned boats on the canal are underway to a destination. But there are a few nomadic souls who float along from one free tie-up or backwater anchorage to another. It's still possible, at least for a few

A quiet morning at the town of Clyde's canal landing. *Author collection.*

weeks or months, to live a largely rent-free life on the canal just as the shanty boat people did on the previous towpath version of the Old Erie.

William Dunlop wrote a historical play about people on a packet boat journey across the state called *A Trip to Niagara Falls* shortly after the original Erie Canal's opening that captured the journey's dual nature of tedium and romance. While I find much of interest here, the famous author Nathaniel Hawthorne wrote of his nineteenth-century canal travel on "an interminable mud-puddle" past "dismal swamps." It's all a matter of attitude. If you are open to it, you'll find each day holds a promise of new sights around the next bend. "Ray," the mascot heron who learned to fish in the lock's discharge waters; *Dragonfly*, the home-built solar/diesel hybrid boat; or college crews out with their eights and their coach training for an upcoming regatta competition may enliven the scene. As do dawn's misty waters, summer's thunderheads building lofty overhead and then the stillness of dusk with a waxing first quarter moon reflected in the dark water.

The canal's future is uncertain. If you want to try canaling with a boat on America's most famous and historic artificial waterway, don't wait too long.

BIBLIOGRAPHY

Adams, Samuel Hopkins. *The Erie Canal.* New York: Random House, 1953.
Beebe, Dr. David. *Camillus Half Way There.* Self-published, 2008.
Campbell, William W. *The Life and Writings of DeWitt Clinton.* https://www.eriecanal.org/texts/Campbell/chap06-1.html.
Doyle, Michael. *The Forestport Breaks: A Nineteenth Century Conspiracy Along the Black River Canal.* Syracuse, NY: Syracuse University Press. 2004.
Edmonds, Walter D. *Rome Haul.* Boston: Little and Brown, 1929.
The Erie Canal. https://www.eriecanal.org/links.html.
Erie Canalway. https://eriecanalway.org/explore/boating/tours-rentals.
Garrity, Richard. *Canal Boatman.* Syracuse, NY: Syracuse University Press, 1977.
Kelly, Jack. *Heaven's Ditch: God, Gold and Murder on the Erie Canal.* New York: St. Martin's Press, 2016.
McFee, Michele A. *A Long Haul: The Story of the New York State Barge Canal.* Fleischmanns, NY: Purple Mountain Press, 1998.
New York State Canal Recreationway Plan. https://www.canals.ny.gov/news/crc/c4.pdf.
O'Mally, Charles T. *Low Bridges and High Water on the New York State Barge Canal.* Utica, NY: North Country Books, 1991.
Rapp, Marvin A. *Canal Water and Whiskey Tall Tales of the Erie County.* Buffalo, NY: Canisius College, 1992.

Riley, Mike. "The Cat Tail Company: Montezuma Fiber." New York Almanack. https://www.newyorkalmanack.com/2013/10/the-cat-tail-company-montezuma-fibre/.

Stack, D.D., and R.S. Marquisee. *The Erie Canal.* Media Artists Inc., 2001.

Whitford, Noble E. *History of the Barge Canal of New York State*. 1923. https://www.eriecanal.org/texts/Whitford/1906/contents.html.

ABOUT THE AUTHOR

Susan Peterson Gateley is a native Upstate New Yorker, former teacher and ichthyoplankton taxonomist, who single-handed the wooden sloop *Ariel* in search of stories for boating and nature magazines for seventeen years. In 1996, she acquired a different boat and partner and added Gateley to her name. Her articles about upstate waters have run in numerous local and national publications, and she has authored a half-dozen titles on Lake Ontario. She has a master's degree in fisheries science and has sailed coastal waters, Lake Ontario and, more recently, the New York Canal System, with vessels ranging from fourteen to forty-seven feet. Visit susanpgateley.com for more on her titles and a link to her Substack newsletter, *The Lake Ontario Chronicles*.

Books by the Author

Order from susanpgateley.com

Ariadne's Death: Tales of Heroism and Tragedy on Lake Ontario
Descriptions of some of the most dramatic shipwrecks, exciting rescues and really close calls on Lake Ontario between 1840 and 2002.

Legends and Lore of Lake Ontario (Arcadia Press, 2013)
Lake monster sightings, ghost stories, smuggling and salvage tales along with those of three fish are included in this collection of lake-related lore.

Living on the Edge with Sara B
When an unexpected inheritance lands in the author's mailbox and is spent on eBay for a small, elderly schooner, a journey ensues and the adventure begins.

Maritime Tales of Lake Ontario (Arcadia Press, 2012)
Shipwrecks and survival on an inland sea.

A Natural History of Lake Ontario (Arcadia Press, 2021)
Localized lake weather, geology, seasonal changes, fisheries, beach dynamics and more for the lake's sailors, beachcombers and nature watchers—includes many illustrations by the author.

Passages on Inland Waters
Jesuit journeys to save souls, a widow's wilderness flight, sailing through Lake Ontario's perfect storm and a journey on the old Erie Canal are among the varied passages you'll find in this collection of historic and modern-day voyages made aboard a variety of vessels.

Saving the Beautiful Lake: A Quest for Hope
A recent environmental history of Lake Ontario with photos.

Twinkle Toes and the Riddle of the Lake
A story for the young and the young at heart in which a crabby cat, a lousy navigator and an old wooden boat journey to Canada to find answers to a mysterious disappearance.

The Widow Maker: A Maritime Tale of Lake Ontario
In 1880, Mollie McIntyre, recently widowed and now owner of the family-operated cargo schooner *Gazelle*, fights two battles to keep her business afloat: one with Lake Ontario, the other with the male-dominated waterfront of her day.